T0391388

Pattern in Music

This book presents analyses of pattern in music from different computational and mathematical perspectives.

A central purpose of music analysis is to represent, discover, and evaluate repeated structures within single pieces or within larger corpora of related pieces. In the chapters of this book, music corpora are structured as monophonic melodies, polyphony, or chord sequences. Patterns are represented either extensionally as locations of pattern occurrences in the music or intensionally as sequences of pitch or chord features, rhythmic profiles, geometric point sets, and logical expressions. The chapters cover both deductive analysis, where music is queried for occurrences of a known pattern, and inductive analysis, where patterns are found using pattern discovery algorithms. Results are evaluated using a variety of methods including visualization, contrasting corpus analysis, and reference to known and expected patterns.

Pattern in Music will be a key resource for academics, researchers, and advanced students of music, musicology, music analyses, mathematical music theory, computational musicology, and music informatics. This book was originally published as a special issue of the *Journal of Mathematics and Music*.

Darrell Conklin is an Ikerbasque Research Professor in the Department of Computer Science and Artificial Intelligence at the University of the Basque Country, San Sebastian, Spain.

Pattern in Music

Edited by

Darrell Conklin

CRC Press is an imprint of the
Taylor & Francis Group, an **informa** business

First edition published 2024
by CRC Press
4 Park Square, Milton Park, Abingdon, Oxon, OX14 4RN

and by CRC Press
2385 NW Executive Center Drive, Suite 320, Boca Raton FL 33431

CRC Press is an imprint of Informa UK Limited

British Library Cataloguing-in-Publication Data
A catalogue record for this book is available from the British Library

ISBN: 978-1-032-60094-9 (hbk)
ISBN: 978-1-032-60095-6 (pbk)
ISBN: 978-1-003-45754-1 (ebk)

DOI: 10.4324/9781003457541

Typeset in Times
by codeMantra

Publisher's Note
The publisher accepts responsibility for any inconsistencies that may have arisen during the conversion of this book from journal articles to book chapters, namely the inclusion of journal terminology.

Contents

Citation Information

The chapters in this book were originally published in the *Journal of Mathematics and Music*, volume 15, issue 2 (2021). When citing this material, please use the original page numbering for each article, as follows:

Introduction
Pattern in music
Darrell Conklin
Journal of Mathematics and Music, volume 15, issue 2 (2021) pp. 95–98

Chapter 1
Discovering distorted repeating patterns in polyphonic music through longest increasing subsequences
Antti Laaksonen and Kjell Lemström
Journal of Mathematics and Music, volume 15, issue 2 (2021) pp. 99–111

Chapter 2
Mining contour sequences for significant closed patterns
Darrell Conklin
Journal of Mathematics and Music, volume 15, issue 2 (2021) pp. 112–124

Chapter 3
Parsimonious graphs for the most common trichords and tetrachords
Luis Nuño
Journal of Mathematics and Music, volume 15, issue 2 (2021) pp. 125–139

Chapter 4
Triadic patterns across classical and popular music corpora: stylistic conventions, or characteristic idioms?
David R. W. Sears and David Forrest
Journal of Mathematics and Music, volume 15, issue 2 (2021) pp. 140–153

Chapter 5
Modelling pattern interestingness in comparative music corpus analysis
Kerstin Neubarth and Darrell Conklin
Journal of Mathematics and Music, volume 15, issue 2 (2021) pp. 154–167

Chapter 6

A computational exploration of melodic patterns in Arab-Andalusian music
Thomas Nuttall, Miguel G. Casado, Andres Ferraro, Darrell Conklin, and Rafael Caro Repetto
Journal of Mathematics and Music, volume 15, issue 2 (2021) pp. 168–180

Chapter 7

Some observations on autocorrelated patterns within computational meter identification
Christopher Wm. White
Journal of Mathematics and Music, volume 15, issue 2 (2021) pp. 181–193

Chapter 8

Exploring annotations for musical pattern discovery gathered with digital annotation tools
Darian Tomašević, Stephan Wells, Iris Yuping Ren, Anja Volk, and Matevž Pesek
Journal of Mathematics and Music, volume 15, issue 2 (2021) pp. 194–207

Notes on Contributors

Rafael Caro Repetto is a Senior Scientist in the Institute for Ethnomusicology at the Universität für Musik und darstellende Kunst Graz, Austria.

Darrell Conklin is an Ikerbasque Research Professor in the Department of Computer Science and Artificial Intelligence at the University of the Basque Country, San Sebastian, Spain.

Andres Ferraro is a Research Scientist in Pandora-SiriusXM, Spain.

David Forrest is an Associate Professor of Music Theory at Texas Tech University, USA.

Miguel G. Casado is a Software Engineer at Dolby Laboratories, Barcelona, Spain.

Antti Laaksonen is a University Lecturer in Computer Science at the University of Helsinki, Finland.

Kjell Lemström is a Senior University Lecturer in Computer Science at the University of Helsinki, Finland.

Kerstin Neubarth† was a Development Manager for European and International Affairs at the Europa-Universität Flensburg, Germany.

Luis Nuño is a Professor in the Department of Communications at the Polytechnic University of Valencia, Spain.

Thomas Nuttall is a PhD Candidate at the Universitat Pompeu Fabra, Barcelona, Spain.

Matevž Pesek is an Assistant Professor at the Faculty of Computer and Information Science at the University of Ljubljana, Slovenia.

Iris Yuping Ren is a PhD Candidate in the Department of Information and Computing Sciences at Utrecht University, the Netherlands.

David R. W. Sears is an Assistant Professor of Interdisciplinary Arts at Texas Tech University, USA.

Darian Tomašević is a Young Researcher and PhD Student at the Faculty of Computer and Information Science at the University of Ljubljana, Slovenia.

Anja Volk is a Professor of Music Information Computing in the Department of Information and Computing Sciences at Utrecht University, the Netherlands.

Stephan Wells is a Postgraduate Student in the Department of Information and Computing Sciences at Utrecht University, the Netherlands.

Christopher Wm. White is an Associate Professor of Music Theory at the University of Massachusetts, Amherst, USA.

Introduction: Pattern in music

Pattern in music, referring to the discovery, representation, selection, and interpretation of repeated structures within single pieces (*intra-opus*) or corpora (*inter-opus*), is a central part of music analysis, musical style and genre, improvisation, music perception, and composition. This special issue of the *Journal of Mathematics and Music* presents a diverse selection of papers on the topic of pattern in music from computational and mathematical perspectives. The following overview will introduce the papers considering three facets: representation, discovery, and evaluation and interpretation.

1. Representation

The basis of any study of pattern in music is the type of music structure under consideration: in this issue, chord sequences (Nuño; Sears and Forrest); melodies (Conklin; Neubarth and Conklin; Nuttall et al.; Tomašević et al.); and polyphony (Laaksonen and Lemström; White).

Patterns can be represented either *extensionally* as locations of pattern occurrences in the music (Tomašević et al.; White) or *intensionally* as a compact description that is satisfied by sequences of music objects (Conklin; Laaksonen and Lemström; Neubarth and Conklin; Nuño; Nuttall et al.; Sears and Forrest).

For intentional representations, it is necessary to define the syntax of patterns, and the semantics outlining precisely what music fragments are matched. In general, one can make a distinction between *relational* and *feature* representations. Relations are defined over pairs of basic music objects, while features are attribute-value tuples describing music objects viewed in isolation.

Relational

Nuño considers, for trichords and tetrachords, their single-semitone transformations (changing one note in a chord by a semitone). These transformations are applied to all chords over a chord vocabulary to form visualizations referred to as *cyclopes*.

Laaksonen and Lemström consider *point set* patterns: vectors relating multiple points in a two-dimensional onset-pitch space, and pattern occurrences are related by rigid transposition. Via a representation change the authors show how these patterns can be represented as *sequences* of onset-time-pairs for any fixed transposition interval.

In the paper by Tomašević et al., patterns are annotated in graphical annotation tools; when patterns are internally represented as sequences of intervals, transposed occurrences can be found by pattern matching.

Conklin considers patterns where the successive components describe contour relations between notes. In this case, the notes are unknown and the contour relations are derived directly from manuscripts using adiastematic neume notation. Contour relations do not form a mathematical group structure acting on notes, and occurrences of a pattern cannot in general be derived by transposition.

Features

In two papers, the components of sequential patterns are *features* that do not have an explicit intervallic structure. This type of representation is possible when pieces in a corpus are, for example, in a shared tonality (Nuttall et al.: pitch features), or if all events are related to a tonic, for example, as chord sequences represented by Roman numerals as in the paper by Sears and Forrest.

In the paper by White, polyphonic pieces are vertically sliced and various different attributes (onset, duration, and average pitch) are computed for each slice of notes, thereby creating sequences of features which are compared under different rotations.

Though most papers are concerned with sequential patterns, by widening the idea of pattern to any *inter-opus* recurrence any describable predicate on pieces can be considered a pattern. The paper by Neubarth and Conklin considers patterns described by *global features*: attributes that are defined for an entire piece rather than for events within pieces. In that study, global patterns are naturally extracted from the observations of a musicologist when describing groups of pieces.

2. Discovery

Most papers in this issue perform some form of *inductive* data analysis, where patterns are discovered or revealed in music. Inductive pattern discovery is in some cases from a single piece (*intra-opus* patterns: Conklin; Laaksonen and Lemström; Nuño; White) and in others from a large corpus (*inter-opus* patterns: Neubarth and Conklin; Nuttall et al.; Sears and Forrest). Inductive analysis can be contrasted with *deductive* analysis, where a corpus or a single piece is queried for occurrences of a known or annotated pattern (Neubarth and Conklin; Tomašević et al.).

In the several papers concerned with sequential patterns, the methods used for discovery vary widely: using or adapting algorithms for mining sequential patterns (Conklin), and more specifically in natural language text (Laaksonen and Lemström; Nuttall et al.; Sears and Forrest). Neubarth and Conklin apply association rule mining over global features of songs.

A distinction that can be made for pattern discovery algorithms is whether an inductive method produces the entire set of representable patterns, which are later filtered, or whether the visited set of patterns is somehow reduced. Constrained pattern discovery is especially necessary when the virtual set of possible patterns is too large to efficiently enumerate. Exhaustive methods include n-gram enumeration (Nuttall et al.; Sears and Forrest), while Conklin constrains the search to *closed* patterns: those not contained in any longer pattern with the same count in the corpus. Laaksonen and Lemström constrain the search, from the space of all patterns, to the *maximal* patterns ending at each point in the piece.

3. Evaluation and interpretation

For inductive data analysis, a large number of patterns can exist in a single piece or a corpus, and it is important to filter out uninteresting patterns and rank the possibly interesting ones, at some stage of the analysis process.

Visualization

In the paper by Nuño, patterns are geometric traces of considered pieces through the predetermined transformation network. Thus the structure of the transformation space between chords provides a bias to the analyst for which patterns are elegant and interesting. In the paper by White, patterns of self-similar segments in the analysis score are apparent as peaks in autocorrelation within several possible temporal rotations of the music surface.

Contrast analysis

Three papers in this issue perform *contrast analysis*, where a corpus is partitioned and one group is contrasted with another in order to find significant differences that can be described by patterns.

In the paper by Nuttall et al., transcriptions of Arab-Andalusian music are partitioned into classes, each forming in turn an analysis group which is contrasted with the remaining transcriptions. The authors consider measures from natural language corpus mining for revealing stylistic building blocks.

Sears and Forrest consider patterns as short contiguous sequences of Roman numerals in corpora described by chord sequences. The corpus is partitioned into two groups: popular and classical. Two fundamentally different ways are used to measure pattern interest: either by contrasting one set against another or analytically by considering how patterns might occur in random sequence data drawn from the same population.

Neubarth and Conklin partition songs into two groups – older and more recent Teton Sioux songs – following the annotations provided by musicologist Frances Densmore in one of the monographs written during her long career in transcribing Native American music. Patterns are evaluated using measures previously defined for association rule mining.

Reference patterns

A further method for evaluating patterns relies on existing *reference patterns*, for example, in scores that have been independently annotated for patterns or derived from existing musicological writings.

Given a set of reference patterns, pattern induction methods can be evaluated by precision and recall of discovered patterns against reference patterns. Nuttall et al. use a reference set of *centones* – building blocks of melodic mode – provided by a prior musicological analysis. Three different pattern induction methods are evaluated to explore the feasibility of recalling these patterns. While many reference patterns can be retrieved, there also exists a subset that cannot be found by any of the explored methods. This highlights the fact that repetition of patterns is not the only factor of musical salience.

Conklin defines a statistical significance measure for intra-opus patterns and uses this to evaluate closed, maximal, and minimal patterns discovered in Mozarabic chant. Different pattern sets are evaluated against reference patterns called *formulae*: short repeated patterns within the chant. Good precision and recall with respect to the reference patterns can be achieved by considering statistically significant patterns.

In the paper by Neubarth and Conklin, reference patterns are assigned a relative interest level, based on the musicological description by Frances Densmore, and this assignment gives an ordered set of reference patterns to study statistical measures for pattern ranking. Some pattern ranking measures closely follow the relative interest level suggested by Densmore's analysis.

Creating reference patterns is a challenge requiring a deep commitment from musicologists or listeners. Tomašević et al. report on two tools for pattern annotation in piano roll and score notation and study the inter-annotator agreement in practical annotation tasks. Such tools hold promise for providing diverse sets of reference patterns for evaluation of inductive pattern discovery.

Acknowledgments

This special issue of the *Journal of Mathematics and Music* was made possible by the enthusiastic support of the journal editors Jason Yust and Emmanuel Amiot, who also handled the editorial process for papers co-authored by the Guest Editor. It would not have been possible without the help and dedication of the many experts in the areas of mathematics, computation, and music, who reviewed manuscripts. Finally, of course, thanks to all authors who have shared the results of their research in these papers.

Darrell Conklin 

Disclosure statement

No potential conflict of interest was reported by the author(s).

References

Conklin, Darrell. 2021. “Mining Contour Sequences for Significant Closed Patterns.” *Journal of Mathematics and Music* 15 (2): 112–124.

Laaksonen, Antti, and Kjell Lemström. 2021. “Discovering Distorted Repeating Patterns in Polyphonic Music Through Longest Increasing Subsequences.” *Journal of Mathematics and Music* 15 (2): 99–111.

Neubarth, Kerstin, and Darrell Conklin. 2021. “Modelling Pattern Interestingness in Comparative Music Corpus Analysis.” *Journal of Mathematics and Music* 15 (2): 154–167.

Nuño, Luis. 2021. “Parsimonious Graphs for the Most Common Trichords and Tetrachords.” *Journal of Mathematics and Music* 15 (2): 125–139.

Nuttall, Thomas, Miguel G. Casado, Andres Ferraro, Darrell Conklin, and Rafael Caro Repetto. 2021. “A Computational Exploration of Melodic Patterns in Arab-Andalusian Music.” *Journal of Mathematics and Music* 15 (2): 168–180.

Sears, David R. W., and David Forrest. 2021. “Triadic Patterns Across Classical and Popular Music Corpora: Stylistic Conventions, Or Characteristic Idioms?” *Journal of Mathematics and Music* 15 (2): 140–153.

Tomašević, Darian, Stephan Wells, Iris Yuping Ren, Anja Volk, and Matevž Pesek. 2021. “Exploring Annotations for Musical Pattern Discovery Gathered with Digital Annotation Tools.” *Journal of Mathematics and Music* 15 (2): 194–207.

White, Christopher Wm. 2021. “Some Observations on Autocorrelated Patterns Within Computational Meter Identification.” *Journal of Mathematics and Music* 15 (2): 181–193.

Discovering distorted repeating patterns in polyphonic music through longest increasing subsequences

Antti Laaksonen and Kjell Lemström

We study the problem of identifying repetitions under transposition and time-warp invariances in polyphonic symbolic music. Using a novel onset-time-pair representation, we reduce the repeating pattern discovery problem to instances of the classical problem of finding the longest increasing subsequences. The resulting algorithm works in $O(n^2 \log n)$ time where n is the number of notes in a musical work. We also study windowed variants of the problem where onset-time differences between notes are restricted, and show that they can also be solved in $O(n^2 \log n)$ time using the algorithm.

2012 Computing Classification Scheme: music retrieval; pattern matching

1. Introduction

Being able to identify repetitions in music is important for gathering a rich understanding of music (see e.g. Schenker 1954; Lerdahl and Jackendoff 1983; Bent and Drabkin 1987; Temperley 2001) and, also, an elementary task in solving many music-related computational problems. The problem has been widely studied for decades in text and other linear structures. As real music is almost inevitably polyphonic, meaning that a multitude of parallel tones may sound at any time, the corresponding musical task is much more complex. An exhaustive search through all possible sequences of tones would shortly lead to a combinatorial explosion.

In this paper, we consider the problem of finding repeating patterns in Western equal tempered polyphonic music. Our algorithms work on the symbolic music domain using a geometric point set representation of musical notes. By using a symbolic representation we avoid the very challenging problem of the fundamental frequency estimation of a general polyphonic audio music case. However, with imaginative modifications, algorithms developed to one musical domain may become applicable to the other (see e.g. Laaksonen and Lemström 2013).

It is not only polyphony that makes the music pattern discovery task a hard one. There are many possible sources for distortions that should be taken into account. Distortions may be systematic, for instance, identical patterns may appear in different musical key and/or in different written tempo. Such distortions are easy to handle by looking at relative values instead of absolute values: in pitch at the pitch difference, in duration at the quotient between corresponding durations. Such algorithms are called transposition and time-scale *invariant*, respectively, and there are various systematic distortions that may be overcome by applying an appropriate invariance. Unfortunately, this comes with a price: the more the algorithms allow distortions, the more they also generate false-positive repetitions. Therefore, a post-processing phase is often needed to discard extraneous ones.

Lemström and Wiggings (2009) formalized and gave a sparse invariance taxonomy for music information retrieval (MIR) tasks. The taxonomy considers the most common and relevant MIR features: pitch and onset time. Invariances for a feature can be ordered based on their strength. For instance, invariances for onset information in an increasing order are time-position (allows linear shifts in time), time-scale (allows constant scaling in time) and time-warp (allows order preserving local scaling in time).

It would be useful to be able to pick up a suitable method for the task just by analyzing the nature of the underlying dataset (what kind of systematic distortions there are) for understanding the required invariance combination. For instance, assuming that the transposition invariance is indispensable for a musical pattern matching task, depending on how strong invariance is required, there are already efficient algorithms for the transposition and time-shifting invariance-combination (Ukkonen, Lemström, and Mäkinen 2003; Romming and Selfridge–field 2007), transposition and time-scaling combination (Romming and Selfridge-field 2007; Lemström 2010) and transposition and time-warp combination (Lemström and Laitinen 2011). For the repeating pattern discovery problem, there is a method for the transposition and time-shifting invariance-combination (Meredith, Lemström, and Wiggins 2002) but no methods for the transposition and time-scaling or for the transposition and time-warp combination. In this paper, we present an efficient algorithm for the combination of transposition and time-warp invariance for the repeating pattern discovery problem.

In a typical repeating pattern discovery case, the data set is generated by a conversion from either sheet music or a live performance played by some instruments. The latter setting generates much more distortions, thus a stronger invariance combination is needed to deal with the inherent, systematic distortions. For the pitch feature the transposition invariance is strong enough assuming that the pitches are mostly perfectly played. However, in live performances, the tempo varies and onset times are rarely accurate making the time-warp invariance desirable for the onset times. Moreover, the data sets are expected to be dense, that is, there are many musical events irrelevant to the repetition that occur within the time frame of a repetition. Therefore, we should be able to allow *gaps* in repetitions.

Traditional methods for symbolic music information retrieval (MIR) problems have been based on an approximate string matching framework using edit distance (see e.g. Mongeau and Sankoff 1990; Ghias et al. 1995; Uitdenbogerd and Zobel 2002; Clifford et al. 2006; Typke 2007). As the framework has originally been developed for linear strings, it is straightforwardly applicable to handle monophonic music. In order to be able to cope with polyphonic music with a multitude of simultaneous notes and parallelly developing musical themes, Meredith, Lemström, and Wiggins (2002) suggested a piano-roll-like geometric representation of music where each note is represented by a point in the plane. In this representation, the horizontal location (x-axis) gives the onset time for a musical event and the vertical location (y-axis) its pitch level.

In Figure 1, we both illustrate the geometric representation and give an example of a repeating pattern in polyphonic music. In this example, there is a musical motif played in two different keys and tempos. Note that in this example, the recurrent motif is a time-scaled repetition of the original motif, which also means that it is a time-warped repetition. Interestingly, it seems that finding only time-scaled repetitions is much harder than finding time-warped repetitions, as discussed in Section 4.

Given a set of n notes (a musical work), Meredith et al. considered the problems of discovering all maximal repeating patterns and all their occurrences under transposition and time-shifting invariances (here *maximal* means that we cannot add any notes to the repeating pattern). They presented the algorithms SIA and SIATEC for computing the solutions in $O(n^2 \log n)$ and $O(n^3)$

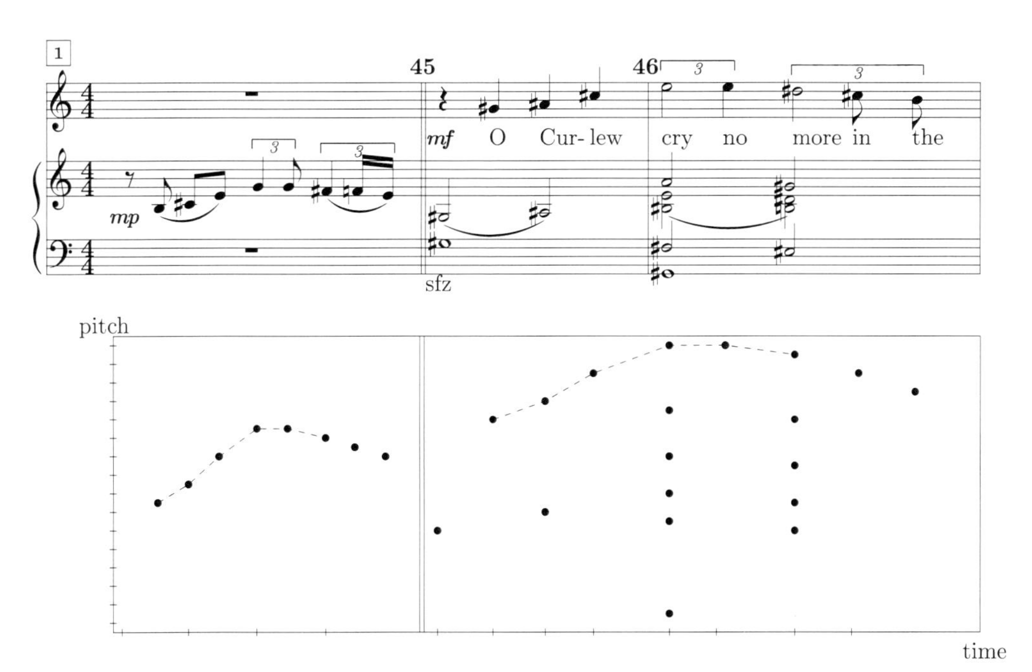

Figure 1. Two excerpts from the piano reduction of Peter Warlock's *The Curlew* (measures 1 and 45–46) in common music notation (above) and the corresponding geometric representation (below) where each note is represented as a point whose x coordinate is the onset time and y coordinate is the pitch. The two representations are synchronized, i.e. the points below are matched with the corresponding notes above, leading to a time-warped point set-representation (note the uneven division in the x axis.) There are several maximal repeating patterns in the example, the most notable being the pattern of six notes represented with dashed lines.

time, respectively. More precisely, the algorithms can be used to process k-dimensional data sets in $O(kn^2 \log n)$ and $O(kn^3)$ time, respectively. In this paper, we only focus on the usual two-dimensional case and ignore the additional k parameter.

The SIA and SIATEC algorithms are tolerant to distortion in the time dimension only to a some extent: if a note is out of time, it is simply discarded. This works rather nicely when only some sporadic notes are distorted. However, when the input is a transcription of a live performance, for instance, the method omits totally the vast majority of the musically meaningful repetitions because a sufficient count of matching individual notes to form a repetition will not be found.

Meredith et al.'s representation gives us a good starting point: it inherently supports point translations (giving us invariances under transposition and pattern location) and sporadic noise omitting. We have to, however, somehow modify it to support arbitrary, local stretching in the time dimension. We will do this by using an onset-time-pair representation where we can solve the original problem of identifying musical repetitions under transposition and time-warp invariances by reducing the problem to instances of the longest increasing subsequence problem.

The rest of the paper is organized as follows: Section 2 discusses the longest increasing subsequence problem with two-dimensional point sets. In Section 3, we describe our repeating pattern discovery algorithm and show variants of the algorithm that can be used to exact and windowed pattern search. All variants of the algorithm work in $O(n^2 \log n)$ time. Finally, Section 4 concludes the paper.

2. Longest increasing subsequences

In this section, we discuss the longest increasing subsequence problem in the context of two-dimensional point sets. It turns out that we can reduce the repeating pattern discovery problem to instances of the longest increasing subsequence problem, and we use the techniques presented in this section in our pattern discovery algorithm.

Consider a set S of n points in the two-dimensional plane. Each point p is represented as a pair $(x[p], y[p])$ where $x[p]$ and $y[p]$ are real numbers. A *subsequence* of length k is a sequence of k points $p_1, p_2, \ldots, p_k$ where $x[p_i] < x[p_{i+1}]$ for $i = 1, 2, \ldots, k-1$. A subsequence is *increasing* if $y[p_i] < y[p_{i+1}]$ for $i = 1, 2, \ldots, k-1$. In the *longest increasing subsequence* problem, we want to calculate for each point $p \in S$ a value $\texttt{lis}(p)$: the maximum length of an increasing subsequence whose last point is p.

For example, let us consider the set

$$S = \{(1,2), (3,5), (4,8), (6,7), (7,2)\}.$$

In this case, $\texttt{lis}((1,2)) = 1$, $\texttt{lis}((3,5)) = 2$, $\texttt{lis}((4,8)) = 3$, $\texttt{lis}((6,7)) = 3$, and $\texttt{lis}((7,2)) = 1$. For example, $\texttt{lis}((6,7)) = 3$, because there is an increasing subsequence $(1,2), (3,5), (6,7)$ whose length is 3.

The longest increasing subsequence problem is a classical algorithm design problem (see e.g. Knuth 1973, 5.1.4; Fredman 1975), and there are several ways to solve it efficiently. Next, we discuss a well-known dynamic programming algorithm for the problem, which can be implemented in $O(n \log n)$ time using a range query data structure. This algorithm will be used as a building block in our pattern discovery algorithm.

2.1. *Dynamic programming algorithm*

We can efficiently solve the longest increasing subsequence problem using a dynamic programming algorithm. To simplify the presentation of the algorithm, we assume that each point in S has a distinct x coordinate and each y coordinate is an integer between 1 and m where $m = O(n)$. However, we will later describe how the algorithm can be modified so that there are no such restrictions.

The idea of the algorithm is to go through the points from left to right (in increasing x coordinate order) and calculate the $\texttt{lis}$ value for each point using the previously calculated values. More precisely, the algorithm can be implemented as follows, assuming that the points are sorted by x coordinate:

```
for i ← 1 to n do
  lis[p_i] ← 1
  for j ← 1 to i − 1 do
    if y[p_j] < y[p_i] then
      lis[p_i] ← max(lis[p_i], lis[p_j] + 1)
    end if
  end for
end for
```

When processing a point p_i, the algorithm initially assumes that $\texttt{lis}[p_i] = 1$. After that, the algorithm goes through all previously processed points $p_1, p_2, \ldots, p_{i-1}$ whose $\texttt{lis}$ values are already known. For each such point, the algorithm checks if the corresponding subsequence can be extended by adding the current point p_i to the subsequence. The final $\texttt{lis}$ value is the maximum length of a subsequence that ends at p_i.

For example, when calculating the value $\texttt{lis}((6,7))$ for the input set $\{(1,2), (3,5), (4,8), (6,7), (7,2)\}$, we have already calculated the values $\texttt{lis}((1,2)) = 1$, $\texttt{lis}((3,5)) = 2$ and $\texttt{lis}((4,8)) = 3$. In this case, we can extend the previous subsequences that end at $(1,2)$ and $(3,5)$ because their y values are below 7. The best solution is to extend the subsequence that ends at $(3,5)$, which yields the value $\texttt{lis}((6,7)) = \texttt{lis}((3,5)) + 1 = 3$.

The running time of the above algorithm is $O(n^2)$ because it consists of two loops that go through the input points. Next, we improve the running time of the algorithm by removing the inner loop using a range query data structure.

2.2. *Improving the algorithm using range queries*

To improve the running time of the dynamic programming algorithm, we use a range query data structure that maintains an array of m numbers (where m is the maximum y coordinate) and can efficiently process the following operations:

- `setVal`(k, x): the array value at position k becomes x
- `getMax`(a, b): find the maximum value between array positions a and b

We assume that the array elements are indexed $1, 2, \ldots, m$ and every array value is initially 0. Each array position corresponds to a y value in the point set. Using such a data structure, we can implement the algorithm as follows:

```
for i ← 1 to n do
    lis[p_i] ← getMax(1, y[p_i] − 1) + 1
    setVal(y[p_i], lis[p_i])
end for
```

The range query structure contains for each possible y value the maximum length of an increasing subsequence whose last point has that y value. To calculate a value `lis`$[p_i]$, the algorithm efficiently finds the length of a previous subsequence whose last point has a y value between 1 and $y[p_i] - 1$, and adds one to that value (if $y[p_i] = 1$, we assume that `getMax`$(1, 0) = 0$). After that, the algorithm updates the range query structure so that it can be used in future searches.

Let us consider again our previous example where the input set consists of points $\{(1, 2), (3, 5), (4, 8), (6, 7), (7, 2)\}$. In this case $m = 8$ and the initial range query array is $[0, 0, 0, 0, 0, 0, 0, 0]$. When the algorithm reaches the point $(6, 7)$, it has already calculated the values `lis`$((1, 2)) = 1$, `lis`$((3, 5)) = 2$ and `lis`$((4, 8)) = 3$ and the range query array is $[0, 1, 0, 0, 2, 0, 0, 3]$. Then, to calculate the value of `lis`$((6, 7))$, the algorithm performs a range query `getMax`$(1, 6) = 2$ and correctly detects an increasing subsequence of length 3. After that, the range query array becomes $[0, 1, 0, 0, 2, 0, 3, 3]$.

The efficiency of the algorithm depends on the operations of the range query structure. It turns out that we can implement both `setVal` and `getMax` in $O(\log m)$ time. Since we have assumed that $m = O(n)$, this yields an $O(n \log n)$ time algorithm. More precisely, we can use a segment tree data structure where each leaf has an array value and each internal node has the maximum value in its subtree (de Berg et al. 2010, Chapter 10; Laaksonen 2017, Chapter 9). For example, Figure 2 shows the segment tree that corresponds to the array $[0, 1, 0, 0, 2, 0, 3, 3]$. Since each array value belongs to $O(\log n)$ subtrees and any range can be divided into $O(\log n)$ distinct subranges that correspond to segment tree nodes, both the operations can be implemented in $O(\log n)$ time.

2.3. *Generalizing the algorithm*

So far we have assumed that every point has a distinct x coordinate and every y coordinate is an integer between 1 and m where $m = O(n)$. However, it is possible to implement the algorithm without such restrictions.

First, to support multiple points with the same x coordinate, we can defer the `setVal` operations using a *buffer* of updates. After calculating a `lis` value, we add the point p_i to the buffer instead of directly performing a `setVal` operation. Then, always when the x coordinate of a

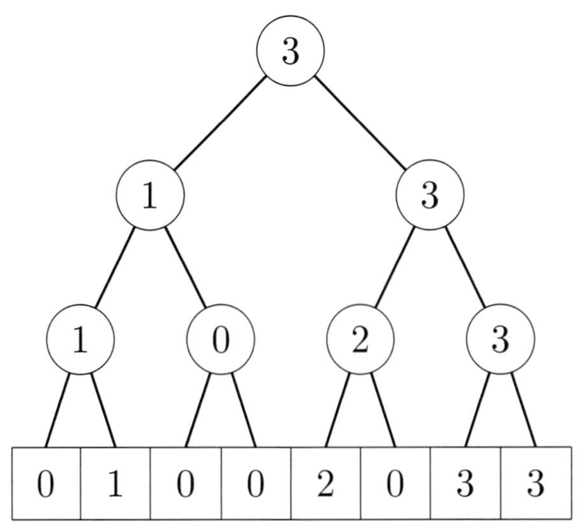

Figure 2. A segment tree that corresponds to the array $[0, 1, 0, 0, 2, 0, 3, 3]$. Using this data structure, we can efficiently calculate the maximum value in an array range and update an array value. Both the operations can be implemented in $O(\log n)$ time.

point is greater than the x coordinate of the previous point, we perform `setVal` operations for all points in the buffer and then clear the buffer. This ensures that there will be no subsequences where two points have the same x coordinate. We can implement the buffer using a stack, and each point will be added to the buffer and removed from the buffer at most once during the algorithm.

Then, to allow arbitrary y coordinates, we can first *compress* the y coordinates so that they become consecutive integers. This can be done by creating a list that contains all y coordinates of the input points and then sorting the list. After that, we can modify the y coordinates so that the smallest coordinate becomes 1, the second smallest coordinate becomes 2, and so on. For example, if the input point set is $\{(1, 500), (4, -5), (5, 8)\}$, it becomes $\{(1, 3), (4, 1), (5, 2)\}$. Since there are at most n distinct coordinates, each coordinate will be an integer between 1 and n after the compression. This works because the algorithm only uses the order of the coordinates and not their exact values.

We can efficiently implement both the generalizations so that the resulting algorithm still works in $O(n \log n)$ time.

3. Finding repeating patterns

In this section, we present our algorithm for finding repeating patterns in polyphonic music. The algorithm solves the pattern discovery problem by reducing it to instances of the longest increasing subsequence problem. Given a musical work represented as a set S of two-dimensional points, the algorithm calculates for each note pair the maximum length of a repetition that ends at those notes.

Each point $p \in S$ corresponds to a musical note. The x coordinate $x[p]$ denotes the onset time of the note, and the y coordinate $y[p]$ denotes the pitch of the note. A repetition of length k consists of two patterns $a_1, a_2, \ldots, a_k \in S$ and $b_1, b_2, \ldots, b_k \in S$ where $x[a_i] < x[a_{i+1}]$, $x[b_i] < x[b_{i+1}]$ and $y[a_{i+1}] - y[a_i] = y[b_{i+1}] - y[b_i]$ for $i = 1, 2, \ldots, k - 1$, i.e. the onset times in both patterns are increasing and each corresponding note pair in the patterns has the same interval.

We first describe the idea of the algorithm in a general setting where the onset-time differences in the patterns are not restricted. After that, we show how the algorithm can be used to find exact repetitions where $x[a_{i+1}] - x[a_i] = x[b_{i+1}] - x[b_i]$ for $i = 1, 2, \ldots, k-1$, creating an alternative to the traditional SIA algorithm. Finally, we present more advanced windowed variants of the algorithm.

3.1. *Reduction to increasing subsequences*

The algorithm is based on two main ideas. First, the algorithm divides the repeating pattern discovery problem into subproblems, each of which has a fixed interval between the two patterns. Second, the algorithm uses an *onset-time-pair representation* to solve each subproblem using the longest increasing subsequence algorithm.

First the algorithm creates a collection of sets $P_0, P_1, P_2, \ldots$ where $P_i = \{(a, b) \mid a \in S, b \in S, y[b] - y[a] = i\}$, i.e. P_i contains all note pairs (a, b) whose interval $y[b] - y[a]$ is constant i. The algorithm processes each set P_i separately and finds all repetitions among each set. When processing a set P_i, the algorithm creates an onset-time-pair representation $C_i = \{(x[a], x[b]) \mid (a, b) \in P_i\}$. Each point in C_i consists of x coordinates of a note pair in P_i. The idea of this representation is that each increasing subsequence in C_i corresponds to a repetition (with interval i) in P_i. The algorithm determines for each point $(x[a], x[b])$ the length of the longest increasing subsequence that ends at that point. This increasing subsequence corresponds to a maximum length repetition in S whose last note pair is (a, b).

As an example, consider a point set

$$S = \{(2,2), (2,4), (2,5), (3,6), (4,4), (5,3), (5,6), (6,1), (6,5)\}.$$

Figure 3(a) shows a repetition of length 3 that consists of patterns $[(2,2), (4,4), (5,3)]$ and $[(2,4), (3,6), (6,5)]$. In this repetition the interval between the patterns is 2. The algorithm detects this repetition when it first creates the set

$$P_2 = \{((2,2), (2,4)), ((2,2), (4,4)), ((2,4), (3,6)), ((2,4), (5,6)), ((4,4), (3,6)), \\ ((4,4), (5,6)), ((5,3), (2,5)), ((5,3), (6,5)), ((6,1), (5,3))\},$$

which consists of all note pairs whose interval is 2, and then the set

$$C_2 = \{(2,2), (2,3), (2,4), (2,5), (4,3), (4,5), (5,2), (5,6), (6,5)\},$$

which contains the onset-time-pair representation of the notes. Figure 3(b) shows the increasing subsequence $[(2,2), (4,3), (5,6)]$ in C_2, which corresponds to the repetition in Figure 3(a). This is a longest increasing subsequence that ends at point $(5,6)$ in C_2, which means that the corresponding repetition is a maximum length repetition with interval 2 that ends at notes $(5,3)$ and $(6,5)$ in S. Note that there are also other longest increasing subsequences that end at point $(5,6)$ in C_2; Table 1 shows all such subsequences and the corresponding repeating patterns.

The total running time of the algorithm is $O(n^2 \log n)$. First, the algorithm can generate the P_i sets in $O(n^2 \log n)$ time, because the total number of note pairs in S is $O(n^2)$ and it is possible to sort the pairs and create the sets in $O(n^2 \log n)$ time. After that, using the longest increasing subsequence techniques discussed in Section 2, the algorithm can process each set P_i in $O(k \log k)$ time, where k denotes the number of note pairs in P_i. Since the sum of all k values is $O(n^2)$, the total time required to process all sets is $O(n^2 \log n)$, regardless of the number of note pairs in each set.

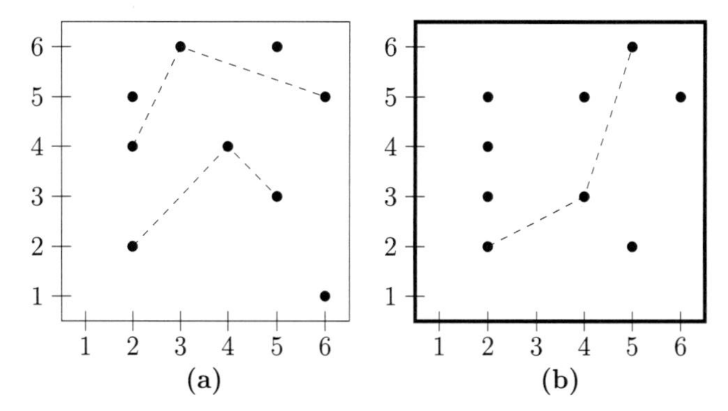

Figure 3. Reduction to the longest increasing subsequence problem for intervals of size two. (a) A repeating pattern that consists of note sequences $[(2,2),(4,4),(5,3)]$ and $[(2,4),(3,6),(6,5)]$. (b) The corresponding longest increasing subsequence $[(2,2),(4,3),(5,6)]$ in the onset-time-pair space.

Table 1. In the example of Figure 3, there are a total of four longest increasing subsequences of length 3 whose last point is $(5,6)$ in the onset-time-pair representation C_2.

Longest increasing subsequence	Repeating pattern occurrence
$[(2,2),(4,3),(5,6)]$	$[(2,2),(4,4),(5,3)]$ and $[(2,4),(3,6),(6,5)]$
$[(2,2),(4,5),(5,6)]$	$[(2,2),(4,4),(5,3)]$ and $[(2,4),(5,6),(6,5)]$
$[(2,3),(4,5),(5,6)]$	$[(2,4),(4,4),(5,3)]$ and $[(3,6),(5,6),(6,5)]$
$[(2,4),(4,5),(5,6)]$	$[(2,2),(4,4),(5,3)]$ and $[(4,4),(5,6),(6,5)]$

Note: This table shows all such subsequences and for each subsequence the corresponding repeating patterns. The first entry in the table corresponds to the repeating pattern shown in Figure 3.

3.2. *Finding exact repetitions*

In the exact version of the problem it is required that $x[a_{i+1}] - x[a_i] = x[b_{i+1}] - x[b_i]$ for $i = 1, 2, \ldots, k-1$. The traditional way to solve the problem is to use the SIA algorithm (Meredith, Lemström, and Wiggins 2002) which first generates and sorts a list of translation vectors between all note pairs in S and then detects repeating patterns as blocks of equivalent vectors in the list. In other words, each block of equivalent vectors corresponds to a maximal translatable pattern of notes. The SIA algorithms works in $O(n^2 \log n)$ time because the list consists of $O(n^2)$ vectors.

We can also approach the exact problem using the longest increasing subsequence technique. In this case, we want to find subsequences whose slope is 1, i.e. the points are located on the same diagonal. For example, Figure 4(a) shows an exact repetition that consists of patterns $[(2,4),(4,4),(5,3)]$ and $[(3,6),(5,6),(6,5)]$, and Figure 4(b) shows the corresponding increasing subsequence $[(2,3),(4,5),(5,6)]$ in C_2. All the points in C_2 are located on the same diagonal.

This variant of the longest increasing subsequence problem is easier to solve than the general problem: we can divide the points in C_i into diagonals where each diagonal has all points (a,b) that have a constant $x[a] - x[b]$ value. For example, in Figure 4(b), all points in the increasing subsequence are located on a diagonal where $x[a] - x[b] = -1$. To find a maximum length repetition, we can simply generate for each diagonal an increasing subsequence that consists of *all* points on the diagonal.

We can go through the diagonals by sorting the pairs by the value of $x[a] - x[b]$, so the algorithm works in $O(n^2 \log n)$ time. In fact, the algorithm works almost like the SIA algorithm: each diagonal in the onset-time-pair representation consists of points where both $x[a] - x[b]$ and $y[a] - y[b]$ are equivalent, which holds exactly when the translation vectors are equivalent.

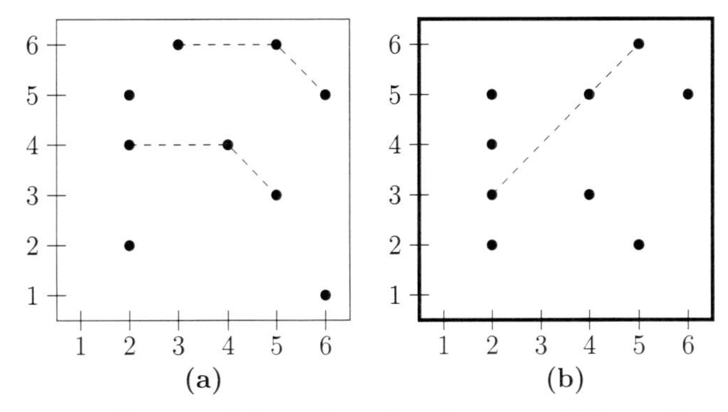

Figure 4. Finding exact repetitions through longest increasing subsequences. (a) A repeating pattern that consists of note sequences $[(2,4),(4,4),(5,3)]$ and $[(3,6),(5,6),(6,5)]$. (b) The corresponding longest increasing subsequence $[(2,3),(4,5),(5,6)]$ in C_2. The repetition is exact, so the slope of the subsequence is 1.

Thus, each diagonal in our algorithm corresponds to a maximal translatable pattern in the SIA algorithm, and the main difference between the algorithms is that our algorithm separately solves a subproblem for each vertical translation (constant interval) while SIA simply processes a sorted list of all vectors.

Note that this exact version of our algorithm finds all maximal translatable patterns, like the SIA algorithm. In other versions of the algorithm, repeating patterns cannot be described using translation vectors because onset-time differences are allowed.

3.3. *Windowed algorithms*

In practice, we often want to discover repeating patterns that have some restrictions in the onset time differences. For example, a pattern where the onset-time difference between two consecutive notes is one minute is probably not musically interesting. One approach for that is to define a window length w and only consider patterns where $x[a_{i+1}] - x[a_i] \leq w$ and $x[b_{i+1}] - x[b_i] \leq w$ for $i = 1, 2, \ldots, k-1$, i.e. the onset-time difference between any two consecutive notes in both patterns is at most w.

It turns out that we can extend our general pattern discovery algorithm so that it supports a window length and still works in $O(n^2 \log n)$ time. To do that, we need a restricted longest increasing subsequence algorithm which ensures that the x and y difference between two consecutive points is at most w. Figure 5 shows the difference between the general and windowed algorithm in the onset-time-pair representation. When we process the point $(6,5)$ and a window length $w = 2$ is used, we can only extend subsequences where x is between 4 and 6 and y is between 3 and 5.

To support the window length in the y dimension, a small modification to the general algorithm is needed. The general algorithm uses the formula `getMax`$(1, y[p_i] - 1) + 1$ to calculate the length of an increasing subsequence that ends at point p_i. Since the first parameter in the range query is 1, any y value below $y[p_i]$ is allowed. Thus, to restrict the y difference, we can modify the query so that the first parameter is $\max(1, y[p_i] - w)$ instead of 1. After this change, the algorithm only considers subsequences where the y difference between two consecutive points is at most w.

The remaining task is to also support the window length in the x dimension. This can be done by removing values from the range query structure after the x coordinate of the corresponding point is no longer inside the window. We can create for each y value an additional data structure that contains all `lis` values of points that are inside the current window and have that y coordinate. The data structure must allow efficient insertion and removal of elements and finding the

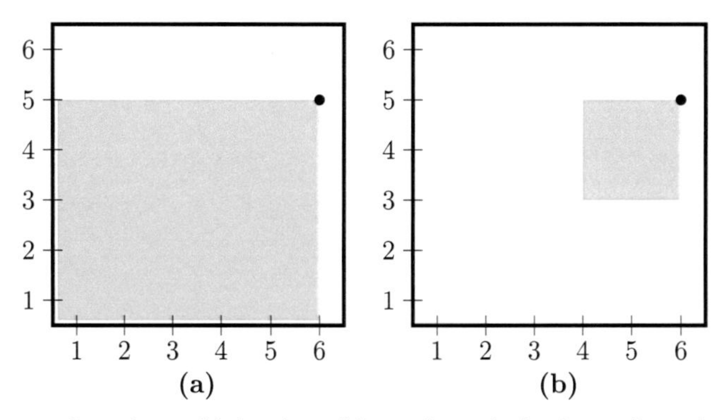

Figure 5. The gray area shows the possible locations of the previous point in a longest increasing subsequence. (a) In the general algorithm, we can choose any point whose x and y coordinates are smaller. (b) In the windowed algorithm, we can only choose points where the x and y difference is at most w. In this example, $w = 2$.

maximum element; we can use a balanced binary search tree that supports all those operations in $O(\log n)$ time. When a point moves outside the window, we first remove the point from the data structure and then update the corresponding value in the range query structure by finding the maximum value for the y coordinate. Since each point is added and removed at most once, the resulting algorithm works in $O(n^2 \log n)$ time like the original algorithm.

Another way to restrict the patterns is to require that the onset-time differences in corresponding pattern positions do not differ too much. This can be done using an additional parameter d and require that $|(x[a_{i+1}] - x[a_i]) - (x[b_{i+1}] - x[b_i])| \leq d$ for $i = 1, 2, \ldots, k-1$, i.e. the onset-time difference is always at most d. For example, if $d = 1$ and the onset-time difference between two notes in the first pattern is 5, then the onset-time difference between the corresponding notes in the second pattern must be between 4 and 6. If $d = 0$, only exact repetitions are allowed.

We can support the d parameter by using two range queries: one in the usual onset-time-pair point set and the other in a *rotated* point set. Figure 6(a) shows the possible point locations using the d parameter, and Figure 6(b,c) represents this area as two rectangles. In both cases, we can use the windowed longest increasing subsequence algorithm to find the repetitions (in the second case we rotate the point set by 45 degrees). Both queries can be processed in $O(\log n)$ time using two range query structures, so the resulting algorithm works in $O(n^2 \log n)$ time.

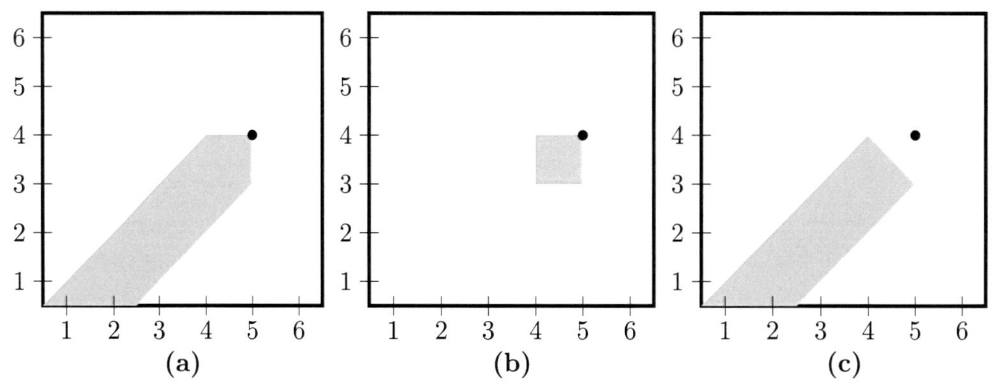

Figure 6. (a) The gray area corresponds to repetitions where the onset-time difference in corresponding pattern positions is at most d (here $d = 1$). (b,c) We can find the repetitions using two windowed queries, one in the original point set and one in a rotated point set.

3.4. *Algorithm implementation*

We have implemented our algorithm in C++, and the source code is available in our GitHub repository (https://github.com/c-brahms/lis-algorithms). The implemented algorithm supports general search without restrictions, exact search and windowed search. The implementation shows that the algorithm works in practice, and it can efficiently process data sets of thousands of notes.

4. Conclusions

In this paper, we have described a new repeating pattern discovery algorithm that can be used to find both exact and time-warped repetitions in $O(n^2 \log n)$ time where n is the number of notes in a musical work. The algorithm is based on an onset-time-pair representation which can be used to reduce the pattern discovery problem to instances of the longest increasing subsequence problem.

In the exact pattern discovery problem, our algorithm can be seen as an alternative (in the two-dimensional case) to the traditional SIA algorithm that also works in $O(n^2 \log n)$ time. The main contribution of our algorithm is that it also solves the more difficult time-warped variant of the problem in $O(n^2 \log n)$ time and supports windows that make the algorithm useful when analyzing real musical inputs.

A general problem in repeating pattern discovery is that many of the detected patterns are usually not musically interesting. This problem can be even more serious in time-warped search because the number of patterns can be large even when the window parameters are well chosen. Thus, in practice, after finding potential repeating patterns, the next step would be to filter musically interesting patterns. In addition, there could be a need for an algorithm like SIATEC that finds all translated occurrences of each maximal translatable pattern. In time-warped search, there are no translation vectors so there cannot be a direct equivalent for SIATEC, but it would still be possible to use a time-warped pattern matching algorithm (Lemström and Laitinen 2011) to find occurrences of each pattern found by the main algorithm.

A future challenge would be to extend the algorithm so that it also supports *time-scaled* repeating pattern discovery. In this variant of the problem, we require that $x[a_{i+1}] - x[a_i] = \alpha |x[b_{i+1}] - x[b_i]|$ for $i = 1, 2, \ldots, k-1$ where α is some constant (each repetition can have an arbitrary α value) which equals the slope of an increasing subsequence (Figure 7 shows

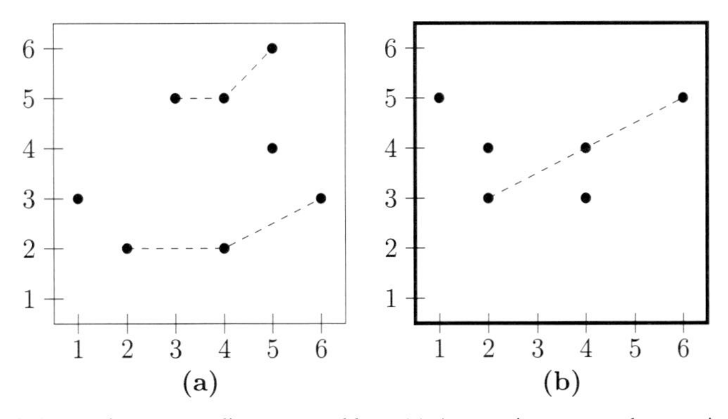

Figure 7. Time-scaled repeating pattern discovery problem. (a) A repeating pattern that consists of note sequences $[(3,5),(4,5),(5,6)]$ and $[(2,2),(4,2),(6,3)]$. (b) The corresponding longest increasing subsequence $[(2,3),(4,5),(5,6)]$ in C_3. The slope of the subsequence equals the scaling parameter $\alpha = 1/2$.

an example where $\alpha = 1/2$). Note that time-scaled and time-warped problems are very different problems from the viewpoint of algorithm design. Time-warped algorithms are usually based on dynamic programming, but this approach is not possible in time-scaled problems and they seem to be more difficult: the best known algorithms for the easier time-scaled pattern matching problem (Lemström 2010) already require quadratic time. Using the longest increasing subsequence technique, we can solve the time-scaled repeating pattern discovery problem in $O(n^4 \log n)$ time by going through all $O(n^2)$ possible α values and performing an individual $O(n^2 \log n)$ search for each of them. However, such an algorithm would not be efficient enough to be used with real musical inputs, so a better approach would be needed to support this invariance.

Disclosure statement

No potential conflict of interest was reported by the author(s).

References

Bent, I., and W. Drabkin. 1987. *Analysis. New Grove Handbooks in Music*. London, UK: Macmillan Press.
Clifford, R., M. Christodoulakis, T. Crawford, D. Meredith, and G. Wiggins. 2006. A Fast, Randomised, Maximal Subset Matching Algorithm for Document-Level Music Retrieval. In *Proceedings of the 7th International Conference on Music Information Retrieval (ISMIR'06)*, 150–155. Victoria, BC, Canada.
de Berg, M., O. Cheong, M. van Kreveld, and M. Overmars. 2010. *Computational Geometry: Algorithms and Applications*. 3rd ed. Berlin, Germany: Springer.
Fredman, M. 1975. "On Computing the Length of Longest Increasing Subsequences." *Discrete Mathematics* 11 (1): 29–35.
Ghias, A., J. Logan, D. Chamberlin, and B. C. Smith. 1995. Query by Humming – Musical Information Retrieval in an Audio Database. In *ACM Multimedia*, 231–236. San Francisco, CA, USA.
Knuth, D. 1973. *The Art of Computer Programming, Volume 3: Sorting and Searching*. Boston, MA, USA: Addison Wesley.
Laaksonen, A. 2017. *Guide to Competitive Programming: Learning and Improving Algorithms Through Contests*. Cham, Switzerland: Springer.
Laaksonen, A., and K. Lemström. 2013. On Finding Symbolic Themes Directly From Audio Using Dynamic Programming. In *Proceedings of the 14th International Society for Music Information Retrieval Conference (ISMIR'13)*, 47–52. Curitiba, Brazil.
Lemström, K. 2010. Towards More Robust Geometric Content-Based Music Retrieval. In *Proceedings of the 11th International Society for Music Information Retrieval Conference (ISMIR'10)*, 577–582. Utrecht, Netherlands.
Lemström, K., and M. Laitinen. 2011. Transposition and Time-Warp Invariant Geometric Music Retrieval Algorithms. In *Proceedings of the 2011 International Conference on Multimedia and Expo (ICME'11)*, 1–6. Barcelona, Spain.
Lemström, K., and G. Wiggins. 2009. Formalizing Invariances for Content-Based Music Retrieval. In *Proceedings of the 10th International Society for Music Information Retrieval Conference (ISMIR'09)*, 591–596. Kobe, Japan.
Lerdahl, F., and R. Jackendoff. 1983. *A Generative Theory of Tonal Music*. Cambridge, MA: MIT Press.
Meredith, D., K. Lemström, and G. Wiggins. 2002. "Algorithms for Discovering Repeated Patterns in Multidimensional Representations of Polyphonic Music." *Journal of New Music Research* 31 (4): 321–345.
Mongeau, M., and D. Sankoff. 1990. "Comparison of Musical Sequences." *Computers and the Humanities* 24(3): 161–175.
Romming, C. A., and E. Selfridge-field. 2007. Algorithms for Polyphonic Music Retrieval: The Hausdorff Metric and Geometric Hashing. In *Proceedings of the 8th International Conference on Music Information Retrieval (ISMIR'07)*, 457–462. Vienna, Austria.
Schenker, H. 1954. *Harmony*. London: University of Chicago Press.
Temperley, D. 2001. *The Cognition of Basic Musical Structures*. Cambridge, MA: MIT Press.

Typke, R. 2007. "Music Retrieval based on Melodic Similarity." PhD thesis, Utrecht University, The Netherlands.

Uitdenbogerd, A., and J. Zobel. 2002. Manipulation of Music for Melody Matching. In *Proceedings of the 6th ACM International Conference on Multimedia*, 235–240. Santa Barbara, CA, USA.

Ukkonen, E., K. Lemström, and V. Mäkinen. 2003. Geometric Algorithms for Transposition Invariant Content-Based Music Retrieval. In *Proceedings of the 4th International Conference on Music Information Retrieval (ISMIR'03)*, 193–199. Baltimore, MD, USA.

Mining contour sequences for significant closed patterns

Darrell Conklin

Sequential pattern mining in music is a central part of automated music analysis and music generation. This paper evaluates sequential pattern mining on a corpus of Mozarabic chant neume sequences that have been annotated by a musicologist with intra-opus patterns. Significant patterns are discovered in three settings: all closed patterns, maximal closed patterns, and minimal closed patterns. Each setting is evaluated against the annotated patterns using the measures of recall and precision. The results indicate that it is possible to retrieve all known patterns with an acceptable precision using significant closed pattern discovery.

1. Introduction

Pattern discovery in music is the process of finding interesting patterns in pieces of music. Music contains patterns at two levels: the *inter-opus* level, occurring in different pieces, and the *intra-opus* level, repeated within single pieces. In music analysis, intra-opus patterns can form the basic structural units that are repeated and transformed throughout a piece. For music generation, pattern discovery has a central role for determining the structure of an existing piece in terms of the units that are repeated. New musical material can be generated to follow the same structure, and this addresses a persistent problem in music generation: modeling long-range dependencies, particularly repetition and reference to earlier generated material.

This paper is concerned with pattern discovery from the perspective of Mozarabic chant. From the 6th until the end of the 11th centuries, Christian worship on the Iberian peninsula was determined by the Mozarabic rite. The chant of this rite is called Old Hispanic chant. Only a few dozen Old Hispanic melodies are known with certainty. However, thousands of melodies of the old Mozarabic rite have been preserved in pitch unreadable, neumatic notation, and the most important collection is the early 10th-century León antiphonary (Randel 1973). Though the pitches of the Old Hispanic melodies are unknown and probably lost forever, the neumes provide important information to assist in their *realization*: determination of a singable and plausible pitch sequence representing the neumes. To achieve this, the neumes in a manuscript can first be transcribed to an abstract contour notation. From note to note it is usually apparent if the melody goes up or down (Rojo and Prado 1929). Figure 1 shows a fragment of the first part of the sacrificium *Aedificavit Moyses* for ordinary Sundays as copied in the León antiphonary (E-L 8, 306r). Shown

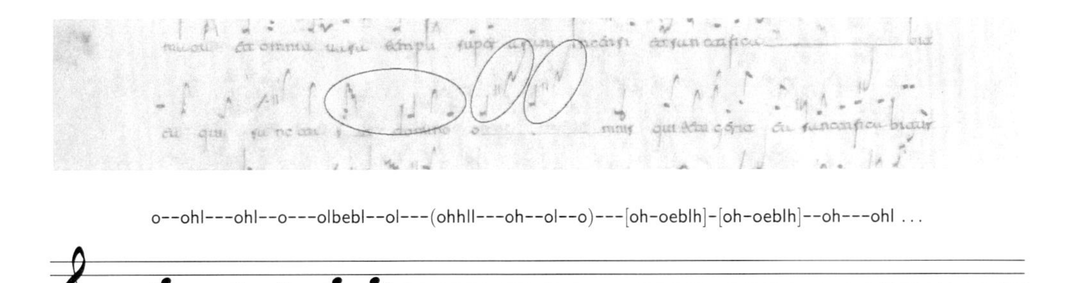

Figure 1. The first part of the sacrificium *Aedificavit Moyses* for ordinary Sundays as copied in the León antiphonary (E-L 8, 306r). In the neumated text (*ea quia sancta sunt domino omnis qui tetigerit ea sanctificabitur*), three instances of two patterns are encircled. Following this is the contour transcription for the first half, with instances of the patterns bracketed. Within the contour sequence, neumes are separated by one dash, syllables by two dashes, and words by three dashes. Bottom: a generated realization of the contour sequence into pitches.

at the top of the figure is one line of text with its neumes from the León antiphonary. Following that is the transcription of the neumes to contour letters. Following Maessen and van Kranenburg (2017), contour information can be represented using six letters (see Figure 1): h, a note higher than the previous note; l, lower; e, equal; b, higher or equal; p, lower or equal; and o, the start of a neume, with an unclear and undefined relative height to the previous note/neume.

In addition to the melodic contour constraints, the melodies often contain *formulae* (intra-opus repeated segments of between approximately 7 to 18 notes) which can be seen in the manuscripts and are retained at an abstract level in the transcribed contour sequences (Maessen and van Kranenburg 2017). In earlier work (Conklin and Maessen 2019), a general realization method was designed for computing melodies that respect both the contour sequence and intra-opus patterns. One assumption made was that all instances of the same intra-opus contour pattern will be instantiated by the same pitches in any realization. Figure 1 (bottom) shows a fragment of a generated chant illustrating respected melodic contour constraints and pitch repetition between intra-opus patterns.

In that earlier work (Conklin and Maessen 2019), patterns were defined through manual annotation of manuscripts. In the extension described in this paper, an automated pattern discovery method is used to find intra-opus patterns. A reliable solution – one that achieves at least high recall of previously annotated chants – could free the musicologist from the intricate task of finding patterns in long neume transcriptions. Additionally, potentially more patterns can be found, including ones that are subtle and perhaps less obvious by visual inspection alone.

This paper develops an intra-opus pattern discovery method and reports on the precision and recall achieved by this method on a corpus of 22 annotated chant templates. The results will also be relevant to any template-based method for music generation (Cope 2001; Collins et al. 2016; Padilla and Conklin 2018), as they indicate how classical concepts of sequential pattern mining can be applied to find structure in an existing piece.

2. Pattern discovery

A *pattern* is a sequence of features representing properties of music events. For the purposes of this paper, referring to Figure 1, a pattern is a sequence of contour letters. An *instance* of a pattern ξ in a sequence is a contiguous subsequence that exactly matches the pattern. The *total count* $n(\xi)$ of a pattern ξ is its total number of instances in the sequence (all overlapping instances are also counted). A pattern ξ is *contained* in a pattern ξ' if $\xi' = X + \xi + Y$ for (possibly empty) patterns X and Y (where $+$ represents the sequence concatenation operator). This relation is written $\xi \sqsubseteq \xi'$. The *length* of a pattern ξ is denoted $|\xi|$.

The fastest algorithms for sequential pattern mining employ a refinement tree search, with each pattern associated with its instance list. Each refinement extends a pattern to the right and produces a more specific pattern (Ayres et al. 2002). At each step a new instance list can be produced rapidly with instance list join operations. All algorithms employ the *downwards closure property* of pattern total count: the total count of a pattern cannot increase as it becomes more specific. This property is invaluable when a minimum threshold on total count is specified, because all refinements of an infrequent pattern are also infrequent patterns.

Even with a high minimum threshold on total count, a huge number of patterns can exist in a given sequence and it is necessary to restrict this space, both structurally (by the types of patterns found) and statistically (as a filter on the pattern set). The first restriction is handled by *closed* patterns, the second by *significant* patterns.

2.1. *Closed patterns*

A structural restriction on patterns is to consider only *closed* patterns (Fournier-Viger et al. 2017). Closed patterns cannot be extended without a drop in the total count. They therefore represent the bottom of containment chains of patterns with the same total count. For inter-opus pattern discovery, usually the shortest, most general patterns are desired (Conklin 2021). On the other hand, for intra-opus pattern discovery, as noted by other researchers (Li, Lee, and Shan 2004; Lartillot 2016; Ren 2016), closed patterns are more appropriate as they cover the most musical material possible without becoming overly specialized.

Definition 2.1 An intra-opus *closed pattern* ξ is a pattern for which there exists no other pattern ξ' such that $\xi \sqsubseteq \xi'$ and $n(\xi) = n(\xi')$.

Thus the closed patterns are the longest patterns among containment chains of patterns with the same count. This set is typically much smaller than the full pattern set and is *lossless*: the full set can be reconstructed from just the closed patterns (Fournier-Viger et al. 2017). Figure 2 presents a small containment hierarchy of patterns, to illustrate the definition of a closed pattern. Every node contains a contour pattern and a hypothetical count. The top node $\top$ is the empty pattern. In the figure, closed patterns are circled and non-closed patterns are boxed.

Several algorithms have been proposed to find all closed patterns in a set of sequences, and all must face the *candidate maintenance problem*: how to determine if an apparently closed pattern is not contained in another closed pattern. The most effective of these algorithms efficiently omit entire paths that cannot possibly lead to closed patterns (Wang, Han, and Li 2007) and thereby eliminate the need for candidate maintenance. This is done by employing the *backscan pruning* method, the effect of which can be seen in the following:

THEOREM 2.1 *For a pattern ξ, if for some feature γ we have $n(\xi) = n(\gamma + \xi)$, then ξ cannot be the prefix of a closed pattern.*

Proof The premise implies that all instances of ξ are preceded by γ. Now assume that some pattern $\xi + \beta$ is closed. Since $\xi + \beta \sqsubseteq \gamma + \xi + \beta$, the closure of $\xi + \beta$ means that $n(\xi + \beta) \neq n(\gamma + \xi + \beta)$, contradicting the assumption. ■

This theorem can have a profound impact on efficiency because if the precondition holds, the entire subtree headed by ξ can be safely pruned from the search space. This is because no

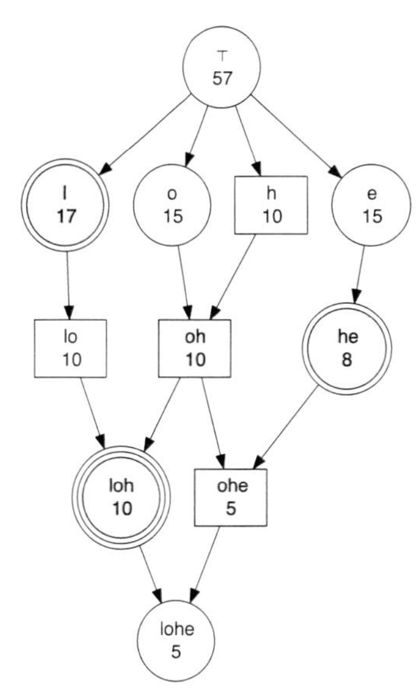

Figure 2. A small containment hierarchy of patterns with their hypothetical counts. Boxed: non-closed patterns; circled: closed patterns; shaded: hypothetical significant patterns; double/triple bordered: minimal and maximal significant closed patterns.

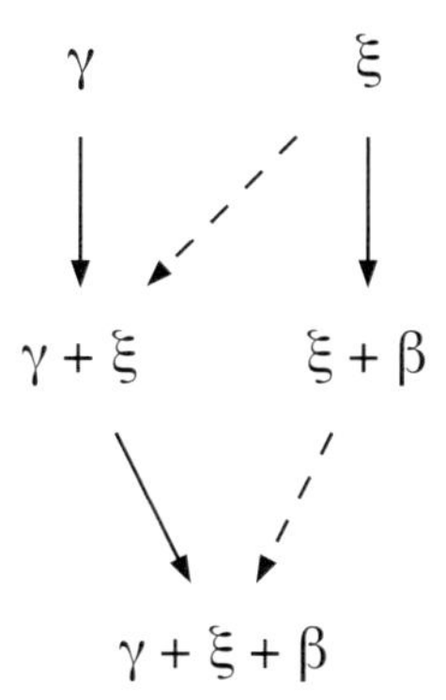

Figure 3. Illustration of Theorem 2.1. Solid arrows represent refinements, and dotted arrows containment relations.

sequence of refinements to the pattern can possibly lead to a closed pattern. Figure 3 illustrates this theorem. If the refinement search sits at ξ, then no refinement of ξ need be applied if $\gamma + \xi$ has the same total count.

Note that the above theorem does not cover the case of minimal closed, or maximal closed patterns, which must be maintained in a candidate list during the search. Referring to Figure 3, if both γ and ξ are closed, since they are on different refinement branches it is still necessary to check for a containment relation between them for either minimal closed or maximal closed pattern mining.

2.2. *Significant patterns*

The set of closed patterns, while smaller than the set of all possible patterns, may still be very large and it is possible to prune these further to patterns that are *significant*: unlikely to occur with the observed count by chance alone. In music simply ranking intra-opus patterns by their count in the sequence is not sufficient, and methods that balance count and prior expectation are necessary (Conklin 2010; Collins et al. 2011). For intra-opus patterns, an expected total count λ can be computed using a background model of contour letters derived from the analysis piece, and a p-value is computed that indicates the probability of at least the observed number of pattern instances in a sequence of the same length.

More precisely, the expected number of instances of a pattern $\xi = c_1, \ldots, c_{|\xi|}$ in a sequence of length ℓ is

$$\lambda = (\ell - |\xi| + 1) \times \prod_{i=1}^{|\xi|} n(c_i)/N \tag{1}$$

where $n(c_i)/N$ is the proportion of contour letter c_i in a corpus of sequences with a total of N events. The expected number of instances λ therefore makes the simplifying assumptions of independence among contour letters, in addition to independence among all $\ell - |\xi| + 1$ possible instances in the sequence.

Let X be a random variable modeling the total count of a pattern ξ, so that $P(X = k)$ is the probability that the observed count is exactly k. From this, the right tail of the Poisson distribution with expected value λ (equation 1) is the probability of finding the pattern k or more times in the sequence of n events:

$$P(X \geq k) = 1 - \sum_{i=0}^{k-1} \text{Poisson}(i; \lambda). \tag{2}$$

With this definition of pattern p-value in hand, three subtypes of closed patterns further reduce the set of possible patterns and will be applied later in this paper:

Definition 2.2 Given a significance level α, a *significant closed pattern* is a closed pattern ξ for which $P(X \geq n(\xi)) \leq \alpha$.

Definition 2.3 A *minimal significant closed pattern* is a significant closed pattern that does not contain any other significant closed pattern.

Definition 2.4 A *maximal significant closed pattern* is a significant closed pattern that is not contained in any other significant closed pattern.

Note in the above definitions: a definition of *minimal significant closed* is the minimal patterns of the significant closed ones, not the significant patterns of the minimal closed ones. Related, the *maximal significant closed* patterns are the maximal of the significant closed patterns, not the significant of the maximal closed ones.

In Figure 2, if the hypothetical significant patterns are shaded, the significant closed patterns are in shaded circles, and the patterns with a doubled circle would be the minimal significant closed patterns and the tripled circle the maximal significant closed pattern.

2.3. *Evaluating pattern discovery*

To evaluate the output of pattern discovery against a known set of patterns, the classic measures of *precision and recall* are used. The evaluation of intra-opus discovered patterns has been considered by Collins (2017) where patterns are defined *extensionally* by instance positions of known (see Figure 1) and discovered patterns in a piece. For the patterns studied in this paper, both discovered and known patterns have *intensional* representations and can be compared directly using sequence comparison.

Allowing for some small mismatches between patterns rather than strict equality, a possible similarity measure can be defined according to $L(\xi_1, \xi_2)$: the length of the *longest semi-global common substring* between two patterns ξ_1 and ξ_2. A semi-global alignment of two strings is one that has no mismatches within the entire region of overlap. This holds when either one string is a substring of the other, or they can be shifted so a prefix of one aligns with a suffix of the other. For example, for the strings eloholhoh and holhoelo the longest common substring is holho, which is not a semi-global common substring because the last character of the first string has a mismatch to the sixth letter of the second string. The longest semi-global common substring is elo which is a prefix of the first string aligned to a suffix of the second.

The similarity between two patterns ξ_1 and ξ_2 can be measured by the Dice coefficient which is normalized to lie between 0 and 1:

$$S(\xi_1, \xi_2) = \frac{2 \times L(\xi_1, \xi_2)}{|\xi_1| + |\xi_2|}. \tag{3}$$

A known pattern π is considered *recalled* or a discovered pattern ξ is considered *precise* if $S(\pi, \xi) \geq \epsilon$. The choice of appropriate ϵ can be guided by a visual inspection of some patterns. From this we chose as a suitable threshold a value of $\epsilon = 0.5$. For the example above, $S(\text{eloholhoh}, \text{holhoelo}) = 6/17$, and therefore these two strings would not be considered similar.

Given one template with known patterns Π and discovered patterns Ξ, the *precision* for the template is the proportion of discovered patterns that are precise, and the *recall* for that template is the proportion of known patterns that are recalled:

$$\text{precision}(\Xi, \Pi) = \left|\{\xi \in \Xi \mid \exists\pi \in \Pi \; S(\xi, \pi) \geq \epsilon\}\right| \, / \, \left|\Xi\right|$$

$$\text{recall}(\Pi, \Xi) = \left|\{\pi \in \Pi \mid \exists\xi \in \Xi \; S(\pi, \xi) \geq \epsilon\}\right| \, / \, \left|\Pi\right|$$

To evaluate the performance over a set of templates, the *mean precision*, and *mean recall* are computed as the means over all templates and their discovered patterns.

2.4. *Chant templates*

An earlier study on the generation of Mozarabic chant (Conklin and Maessen 2019) used 22 different chants each manually annotated with intra-opus patterns. These templates provide an excellent opportunity for evaluating sequential pattern mining algorithms and their different configurations. The templates contain a total of 159 annotated patterns of length 7 or more, in 421 annotated instances. Though the patterns are defined *extensionally* by manual annotation of their boundaries in the contour sequence (see Figure 1), they can also be interpreted *intensionally* because all instances have the same contour sequence. Though it is still possible that an intensionally defined instance is not extensionally defined, only 16 of these patterns show a slight difference between their extensionally annotated count and their intensional count.

Though the contour sequences in the templates are divided into neumes, syllables, and words (see Figure 1) for this study they are viewed simply as unstructured contour sequences. A discovered pattern may begin and end anywhere within a neume, and different instances of patterns can span different types of divisions. This assumption might affect the ability to retrieve annotated patterns exactly, though the effects on recall and precision are mitigated by using inexact matching and permitting some variation between known and discovered patterns (Section 2.3).

The intra-opus patterns fall into three length categories. First, most of the 22 pieces have *repetendae*, longer parts of the chant that should be repeated after a verse. The encoding of the manuscripts copied them always completely, thus creating longer intra-opus patterns. Second, an important feature of chant is the presence of recurring *formulae* which represent the

Table 1. Statistics of 22 templates containing manually annotated intra-opus patterns.

Template	Length	Patterns	Instances	Coverage (notes)
01_SCR012_AL054v10	928	5	11	0.40 (375)
02_RS057_AL066r12	310	1	2	0.61 (188)
03_SCR015_AL072v09	1009	9	30	0.51 (512)
04_SCR023_AL088v08	1258	11	32	0.50 (628)
05_SCR041_AL154v09	852	4	13	0.34 (291)
06_SCR047_AL163v11	1523	11	26	0.32 (481)
07_SCR048_AL175v02	922	11	26	0.56 (520)
08_SCR049_AL177r01	959	12	32	0.52 (494)
09_SCR050_AL177r16	687	6	18	0.58 (401)
10_SCR065_AL210r14	777	6	13	0.45 (346)
11_SNO_AL211v10	731	12	29	0.50 (366)
12_RS291_AL212v02	275	2	5	0.67 (185)
13_RS292_AL212v11	444	2	4	0.31 (138)
14_RS293_AL213r08	206	2	5	0.44 (90)
15_RS294_AL213v01	235	1	2	0.34 (80)
16_RS295_AL213v10	200	1	2	0.40 (80)
17_RS296_AL214r02	365	6	16	0.73 (268)
18_RS297_AL214r12	218	1	2	0.44 (96)
19_VAR_AL214v12	707	8	27	0.46 (325)
20_SCR066_AL215r13	1485	13	33	0.34 (507)
21_SCR069_AL229r14	2569	28	76	0.61 (1573)
22_SCR077_AL240r16	695	7	17	0.40 (275)
mean	789	8	20	0.40

same melodic content in the lost chant, for example, the encircled neumes and bracketed contour sequences in Figure 1. There is a wide consensus among chant scholars that long intra-opus patterns do represent the same sequence of pitches (Hornby and Maloy 2012). Third, very short patterns (length less than 7) are used mainly to indicate repetition within other longer patterns. Only the patterns in the first two categories are considered here.

Table 1 gives some overall statistics of the set of templates, The table shows the number of notes in the template, the number of distinct annotated patterns (of length 7 or more), the number of instances of those patterns, and finally the coverage of the notes of the template by pattern instances. For example, the template 11_SNO_AL211v10 has 731 notes, 12 patterns which have 29 instances, with 50% of the template – 366 notes – covered by those instances. The mean template length is 789 notes, and a mean of 40% of template notes are covered by annotated pattern instances.

3. Results

Intra-opus sequential pattern mining was applied to all 22 templates (Table 1) and this section will report on a detailed analysis for one particular template, followed by a study of precision and recall on all 22 templates. A fixed similarity value (equation 3) of $\epsilon = 0.5$ was used to measure precision and recall. The minimum total count is $n(\cdot) = 2$: patterns must occur at least twice in a sequence. The minimum accepted pattern length was 7. For patterns within a given template, expected pattern counts (equation 1) are computed using a background model of contour letters created from all other templates.

3.1. *Single template*

To illustrate the numbers of different types of patterns discovered, one template 11_SNO_AL211v10 from Table 1 is used. Table 2 summarizes the results. The total set of

Table 2. Numbers of patterns discovered in 11_SNO_AL211v10 at various pattern mining settings (minimum length 7, minimum total count 2).

all patterns	1443
closed patterns	90
minimal closed patterns	47
maximal closed patterns	45
significant ($\alpha = 1e{-}10$) closed patterns	18
minimal significant closed patterns	6
maximal significant closed patterns	7

repeated patterns is large (1443) and closed patterns quickly reduce this set to just 90 patterns. Filtering by statistical significance reduces the set to just 18 patterns. Further reductions are achieved by considering maximal, or minimal, patterns, though as it will be seen both of these reduce precision and recall.

To illustrate the operation of precision and recall, the template has 12 annotated patterns with 29 instances (Figure 4a). The instances of each pattern are positioned along the 731 letters of the template as indicated by horizontal position of bracketed lines; the vertical line indicates the start of the template. The template has several interesting properties that are also present in other templates. Patterns a2, a3, a4, a9, and a12 are *cyclic* (Lartillot 2016): having an immediately contiguous repetition. The template has some hierarchical structure, with patterns a2 and a4 containing repeated subpatterns (patterns a6 and a7), and pattern a3 with an overlapping subpattern (pattern a10, which has an instance in another area of the template).

Figure 4(b) shows the discovered closed patterns for this template at $\alpha = 1e{-}10$. Discovered patterns were sorted by increasing p-value. Regarding the discovered patterns in Figure 4(b), all patterns are precise: they are all similar to one or more annotated patterns. In red in Figure 4(a) are the known patterns that are not recalled in this setting. Thus the recall in this setting is $9/12 = 0.75$. The three missed patterns are somewhat shorter than the mean length, and have several undefined (o) contours, which increases their background probability and therefore their expected count (equation 1). Combined with this, all patterns have only two instances which reduces the left tail of their p-value thereby increasing the right tail (equation 2). These three patterns can be recalled at a more permissive $\alpha = 1e{-}4$, though with a drop in precision to 0.67 (29 rather than 18 closed patterns discovered).

The set of all closed patterns (Figure 4b) contains several patterns (b3, b4, b9, b13, b17) that have some overlap among their instances. For example, pattern b3 has the prefix ohlo which is also a suffix, meaning that the pattern has some *periodicity* and therefore the potential for overlap among its instances. This is manifested in the template 11_SNO_AL211v10, as there is an overlap between the two instances of the pattern.

Figure 4(b) shows that several patterns (b4, b5, b7–b12, b14, and b16) are clustered in the small regions covered by pattern a6 of the template. Pattern b8 (Figure 4b) is an exact match to pattern a6, but several other patterns are found which are very slight variations, each having a different extension and total count varying from just 3 instances (pattern b16) through to 11 (pattern b4). Patterns b4 and b9 indicate a possible short cyclic structure in the template that was not annotated.

Compared with all closed patterns, there are fewer minimal closed patterns (Figure 4c) and maximal closed patterns (Figure 4d). The minimal closed patterns achieve good precision but as expected with fewer and shorter patterns, the recall drops to 0.58. With maximal closed patterns (Figure 4d), the recall increases slightly to 0.67. Neither maximal nor minimal patterns reveal the desired substructure of the template produced by patterns a2 and a3: maximal patterns missing the subpatterns and minimal patterns missing the longer enclosing pattern.

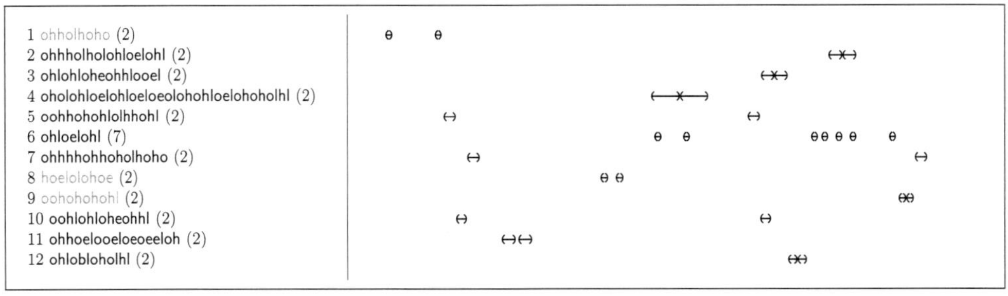

(a) Annotated patterns

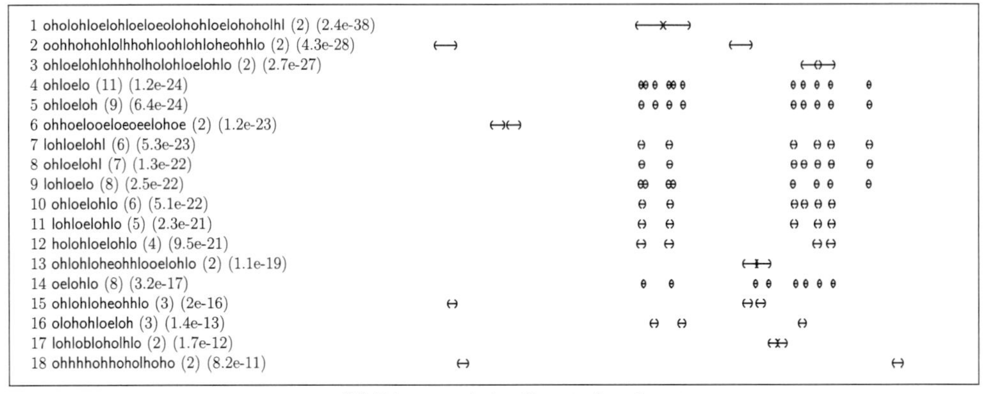

(b) Discovered significant closed patterns

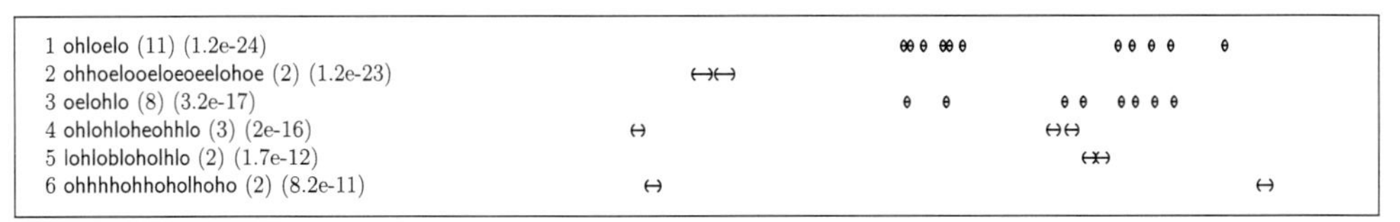

(c) Discovered minimal significant closed patterns

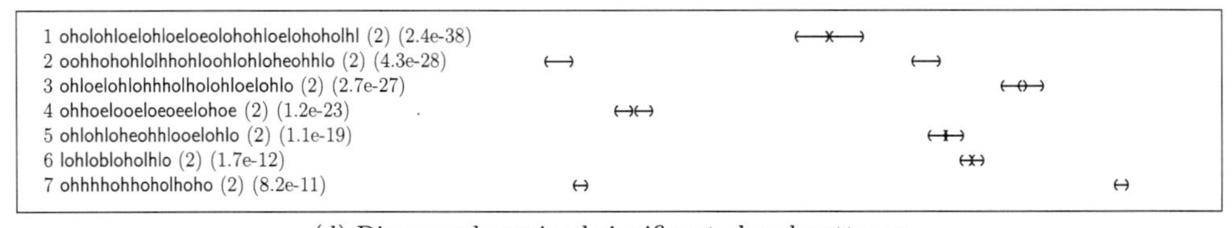

(d) Discovered maximal significant closed patterns

Figure 4. (a) Annotated patterns for template 11_SNO_AL211v10. (b–d) Discovered patterns at $\alpha = 1e{-}10$ for closed (b), minimal (c), and maximal (d) patterns. For discovered patterns, the number of instances and p-values are in brackets following the pattern. (a) Annotated patterns. (b) Discovered significant closed patterns. (c) Discovered minimal significant closed patterns and (d) Discovered maximal significant closed patterns.

This section has studied one template and shown all closed, minimal closed and maximal patterns for one template. For descriptive purposes fewer patterns, as achieved by minimal mining, might be desirable, though for base precision and recall it appears that all closed patterns is the best choice. The next section will deepen this result on 22 templates.

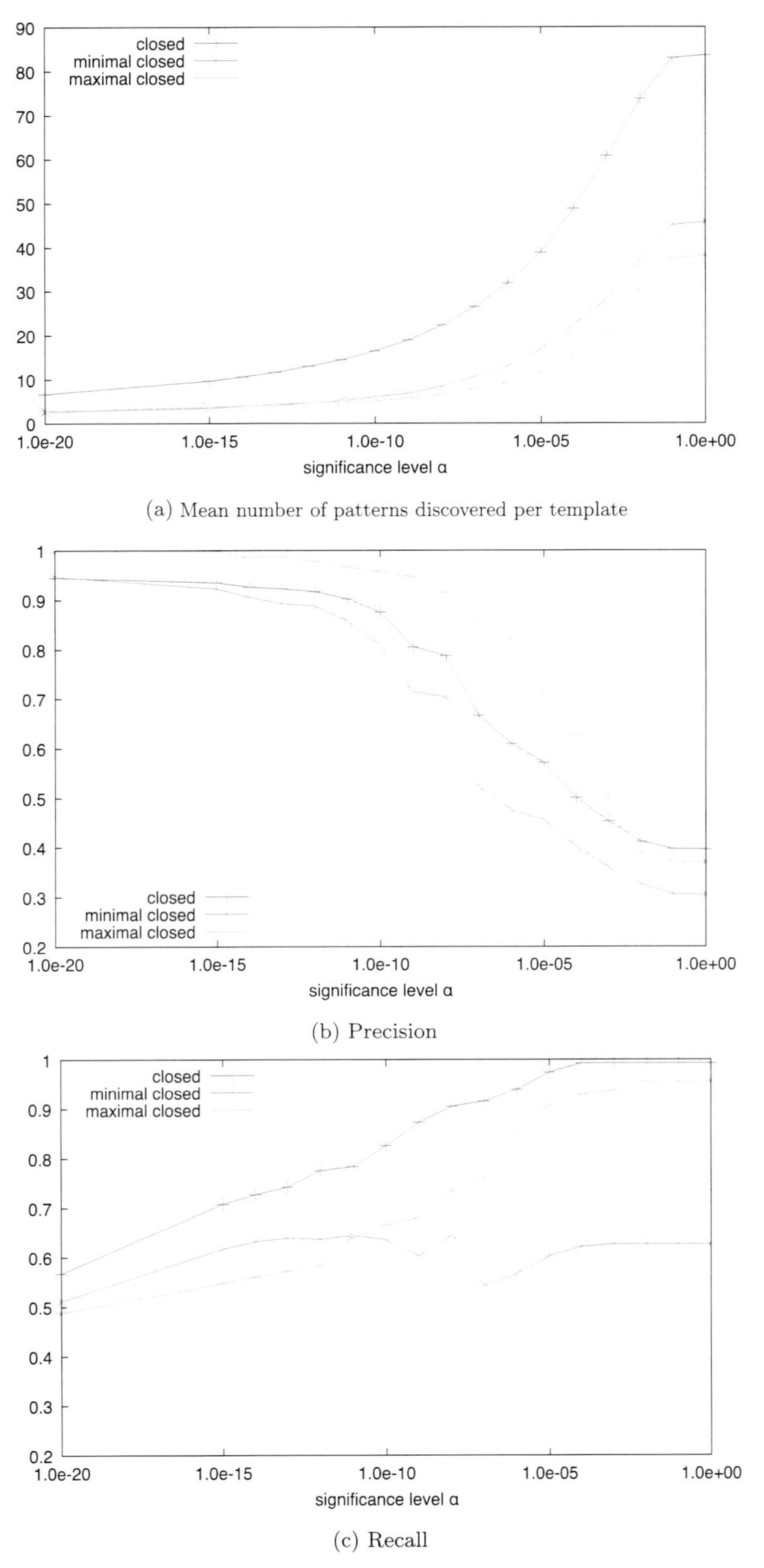

(a) Mean number of patterns discovered per template

(b) Precision

(c) Recall

Figure 5. Number of patterns found, precision, and recall on 22 templates for different pattern discovery settings, varying the significance level α: (a) Mean number of patterns discovered per template, (b) Precision and (c) Recall.

3.2. *Evaluation on 22 templates*

The performance of pattern discovery on all 22 templates was studied by varying the significance level α, the maximum p-value tolerated for discovered patterns (equation 2). Three different settings were considered: all closed patterns, maximal closed patterns, and minimal closed patterns, varying α starting from $1e$–20 through to 1. The results are plotted in Figure 5.

Regarding the number of patterns found (Figure 5a), this increases with increasing α, for all settings, though more slowly for minimal and maximal than for mining all closed patterns. Regarding precision (Figure 5b), as expected this drops as more patterns are discovered. Maximal patterns in general achieve the highest precision over most significance levels. Regarding recall, arguably the most important measure for annotated patterns, all closed patterns have higher recall at all α (Figure 5c). For all settings, recall generally increases for increasing α but reaches a higher level earlier for all closed patterns. Maximal and minimal closed patterns are not competitive for recalling known patterns (Figure 5c). If perfect recall is the goal, then all closed patterns achieve this at $\alpha = 1e$–4, though one must tolerate a loss of precision as about 45 closed patterns per template would be discovered at that level, compared to an average of just 8 annotated patterns per template (Table 1).

For minimal closed patterns, the unexpected behavior of recall (Figure 5c) – sometimes falling with increasing α – can be explained by referring to the definition of minimal patterns. It can happen that a discovered pattern that recalls a known pattern at a given α is not found at a higher α but that a shorter pattern is discovered instead. If the shorter pattern does not recall the known pattern, then a drop in recall might be observed. Though it would be contained in the longer pattern, it might be too short to satisfy the definition of pattern recall at a given ϵ (equation 3).

It was seen above on a single template that mining all closed patterns produces the most patterns, and this carries through to all 22 templates. At a permissible number of roughly 20 discovered patterns per template (see Table 1), at $\alpha = 1e$–8, one would obtain a high recall of ≈ 0.90 by using all closed patterns rather than maximal (≈ 0.70) or minimal (≈ 0.60) patterns. On the other hand, for a low number of patterns per template, with high precision as the goal, maximal patterns are the better setting. If high recall is the goal, with acceptable precision, then for all α settings this is achieved by discovering all closed patterns.

4. Conclusion

Intra-opus pattern discovery in symbolic music data is essential for capturing the meaning of music in a formal way, describing how and where structures repeat throughout a piece (Rolland 1999; Conklin and Anagnostopoulou 2001; Hsu, Liu, and Chen 2001; Cambouropoulos, Crawford, and Iliopoulos 2001; Meredith, Lemström, and Wiggins 2002; Lartillot 2016; Knopke and Jürgensen 2009). Patterns can be used for describing the general semiotic structure of a piece (Ruwet 1966) and for generating new pieces with the same structure (Cope 2001; Collins et al. 2016; Padilla and Conklin 2018; Conklin and Maessen 2019).

In this paper, a set of manually annotated Mozarabic chant templates was used as a basis for the evaluation of pattern discovery under various settings. Attention was given to closed patterns – those at the bottom of containment chains of patterns with the same count – which dramatically reduce the number of discovered patterns. Subsequently, closed patterns were required to be statistically significant: unlikely to occur with their observed count by chance alone. Two further specializations of significant closed patterns were studied: those that are minimal and those that are maximal. The results indicated that maximal patterns have higher precision and lower recall, while minimal patterns are in general not competitive. In a detailed study on one particular chant template, minimal and maximal patterns were unable to recover the detailed known pattern substructure. From the observed high recall of manually annotated patterns, one can conclude that in new unannotated chant templates relevant patterns can be discovered by considering the set of all significant closed patterns.

Acknowledgments

Thanks to Kerstin Neubarth for valuable discussions on the research and the manuscript, and to the reviewers for the helpful comments.

Disclosure statement

No potential conflict of interest was reported by the author.

ORCID

Darrell Conklin http://orcid.org/0000-0002-2313-9326

References

Ayres, J., J. Gehrke, T. Yiu, and J. Flannick. 2002. "Sequential Pattern Mining using A Bitmap Representation." In *Proceedings of the International Conference on Knowledge Discovery and Data Mining*, 429–435. Edmonton, Canada.

Cambouropoulos, E., T. Crawford, and C. S. Iliopoulos. 2001. "Pattern Processing in Melodic Sequences: Challenges, Caveats and Prospects." *Computers and the Humanities* 35 (1): 9–21.

Collins, T. 2017. "MIREX 2017:Discovery of Repeated Themes & Sections." `https://www.music-ir.org/mirex/wiki/2017:Discovery_of_Repeated_Themes_&_Sections`.

Collins, Tom, Robin Laney, Alistair Willis, and Paul H. Garthwaite. 2011. "Modeling Pattern Importance in Chopin's Mazurkas." *Music Perception: An Interdisciplinary Journal* 28 (4): 387–414.

Collins, T., R. Laney, A. Willis, and P. H. Garthwaite. 2016. "Developing and Evaluating Computational Models of Musical Style." *Artificial Intelligence for Engineering Design, Analysis and Manufacturing* 30 (1): 16–43.

Conklin, D. 2010. "Discovery of Distinctive Patterns in Music." *Intelligent Data Analysis* 14 (5): 547–554.

Conklin, D. 2021. "Patterns and antipatterns for music corpus analysis." In *Oxford Handbook of Music and Corpus Studies*, edited by D. Shanahan, A. Burgoyne, and I. Quinn, New York: Oxford University Press.

Conklin, D., and C. Anagnostopoulou. 2001. "Representation and Discovery of Multiple Viewpoint Patterns." In *Proceedings of the International Computer Music Conference (ICMC 2001)*, 479–485. Havana, Cuba.

Conklin, Darrell, and Geert Maessen. 2019. "Generation of Melodies for the Lost Chant of the Mozarabic Rite." *Applied Sciences* 9 (20): 4285.

Cope, D. 2001. *Virtual Music: Computer Synthesis of Musical Style*. Cambridge, MA: The MIT Press.

Fournier-Viger, P., J. C. W. Lin, R. U. Kiran, Y. S. Koh, and R. Thomas. 2017. "A Survey of Sequential Pattern Mining." *Data Science and Pattern Recognition* 1 (1): 54–77.

Hornby, Emma C., and Rebecca Maloy. 2012. "Toward a Methodology for Analyzing the Old Hispanic Responsories." In *Cantus Planus Study Group of the International Musicological Society*, 242–249. Vienna: Österreichische Akademie der Wissenschaften.

Hsu, J-L., C-C. Liu, and A. Chen. 2001. "Discovering Nontrivial Repeating Patterns in Music Data." *IEEE Transactions on Multimedia* 3, 311–325.

Knopke, Ian, and Frauke Jürgensen. 2009. "A System for Identifying Common Melodic Phrases in the Masses of Palestrina." *Journal of New Music Research* 38 (2): 171–181.

Lartillot, O. 2016. "Automated Motivic Analysis: An Exhaustive Approach Based on Closed and Cyclic Pattern Mining in Multidimensional Parametric Spaces." In *Computational Music Analysis*, edited by D. Meredith, 273–302. Cham: Springer.

Li, H. F., S. Y. Lee, and M. K. Shan. 2004. "Mining Frequent Closed Structures in Streaming Melody Sequences." In *2004 IEEE International Conference on Multimedia and Expo (ICME)*, Vol.3, 2031–2034.

Maessen, G., and P. van Kranenburg. 2017. "A Semi-Automatic Method to Produce Singable Melodies for the Lost Chant of the Mozarabic Rite." In *Proceedings of the 7th International Workshop on Folk Music Analysis*, 60–65. Malaga.
Meredith, D., K. Lemström, and G. Wiggins. 2002. "Algorithms for Discovering Repeated Patterns in Multidimensional Representations of Polyphonic Music." *Journal of New Music Research* 31 (4): 321–345.
Padilla, V., and D. Conklin. 2018. "Generation of Two-voice Imitative Counterpoint From Statistical Models." *International Journal of Interactive Multimedia and Artificial Intelligence* 5 (3): 22–32.
Randel, D. 1973. *An Index to the Chant of the Mozarabic Rite*. New Jersey: Princeton University Press.
Ren, I. 2016. "Closed Patterns in Folk Music and Other Genres." In *Proceedings of the 7th International Workshop on Folk Music Analysis*, 56–58. Dublin.
Rojo, C., and G. Prado. 1929. *El Canto Mozárabe, Estudio histórico-critico de su antigüedad y estado actual*. Barcelona: Diputación Provincial de Barcelona.
Rolland, P. Y. 1999. "Discovering Patterns in Musical Sequences." *Journal of New Music Research* 28 (4): 334–351.
Ruwet, N. 1966. "Méthodes D'analyse En Musicologie." *Revue Belge de Musicologie* 20, 65–90.
Wang, J., J. Han, and C. Li. 2007. "Frequent Closed Sequence Mining Without Candidate Maintenance." *IEEE Transactions on Knowledge and Data Engineering* 19, 1042–1056.

Parsimonious graphs for the most common trichords and tetrachords

Luis Nuño

Parsimonious transformations are common patterns in different musical styles and eras. In some cases, they can be represented on the Tonnetz, Cube Dance, Power Towers, or the central region of an orbifold, mainly when they only include the most even trichords and tetrachords. In this paper, two novel graphs, called Cyclopes, are presented, which include more than double the number of chord types in previously published graphs, thus allowing to represent a larger musical repertoire in a practical way. Apart from parsimonious transformations, they are also especially suitable for representing trichords a major third apart, tetrachords a minor third apart, and the cadences V7–I(m) and $II^{\emptyset}$–V7–I(m) with major or minor tonic chords. Therefore, they allow to clearly visualize the relationship among the corresponding chords and better understand those patterns, as well as being efficient mnemonic resources, all of which make them useful tools both for music analysis and composition.

2010 Mathematics Subject Classification: 00A65; 97M80

1. Introduction

Among the recurrent and repeated structures in musical compositions are the parsimonious transformations. They have been widely used in such different musical styles and eras as, for example, Classical period, Romanticism, Latin music, and Jazz, thus being well-established patterns in music. Their analysis can be carried out with the *neo-Riemannian theory*, which arose in the 1980s for analyzing some chromatic passages by nineteenth-century composers and is still evolving with the contributions of algebra and geometry. According to Gollin (2005), it is characterized by three elements: mathematical groups of transformations, voice-leading parsimony, and graphical representations. The paradigmatic example corresponds to the *PLR-group*[1] and the *Tonnetz*, although they are limited to major and minor triads.

As a starting point, a primary rule in harmony for connecting chords is the "law of the shortest way" (Schönberg 1983, 39, quoting Bruckner). This means to sustain the common notes and move the others by the smallest possible intervals. In this respect, Douthett and Steinbach (1998) state that two chords with the same cardinality are $P_{m,n}$*-related* if one of them can be transformed into the other by sustaining the common notes and, for the rest of them, moving m by a semitone

[1] P, L and R stand for the basic operations *Parallel*, *Leading-tone exchange*, and *Relative*, which respectively map, for example, C major to C minor, C major to E minor, and C major to A minor; and vice versa.

and n by a whole tone. Then, *parsimony* is a $P_{m,n}$ relation with low values for m and n, normally $m + 2n \leq 2$. The simplest case is $P_{1,0}$, which we will call *single-semitonal* (after Tymoczko 2011). Douthett and Steinbach (1998) also provide several remarkable parsimonious graphs, particularly the *Chicken-Wire Torus* (the dual of the *Tonnetz*) and *Cube Dance* for nearly and most even trichords, respectively, and the *Towers Torus* and *Power Towers* for nearly and most even tetrachords, respectively. Twenty years before, however, Waller (1978) published a torus equivalent to the Chicken-Wire, but which clearly shows its full hexagonal tessellation, as well as all *PL*, *PR*, and – although a bit harder to visualize – *LR* cycles. These and other *PLR* compound operations were later studied extensively by Cohn (1996, 1997, 1998, and, particularly, 2012). A different approach is given by Tymoczko (2006), who provides the full theory for representing all n-note pitch-class sets in the orbifold $\mathbb{T}^n/S_n$, here abbreviated *n-orbifold*, which is a kind of generalized Möbius strip. As well, he represents the 2-orbifold in 2D and part of the 3-orbifold in 3D, before twisting and bending the figures to obtain the real orbifolds. Callender, Quinn, and Tymoczko (2008) provide further representations in this sense. In practice, however, due to the complexity of the spaces, only the central regions of the orbifolds are normally represented.

In this paper, I present a novel chord pattern representation, based on cyclic circular graphs called "Cyclopes," which show broader groups of trichords and tetrachords related by single-semitonal transformations. As well, they provide a wider view around the center of the corresponding orbifolds. Therefore, they allow to represent a greater number of musical works in a practical way and can be used both for music analysis and composition.

The reader is assumed to be familiar with *Forte names* and *set classes* (Forte 1973), also called *chord classes*. Here, the *non-inversionally-symmetrical* ones are split into two *chord types* related by *inversion*, named "a" and "b", in accordance with Nuño (2020). As well, large parts of this study deal with *chord geometry* (Tymoczko 2011) and *most even chord transformations* (Cohn 2012), although the main concepts are explained here.

2. Dyads

Tymoczko (2006) represents the *unordered pairs of pitch classes*, or simply *two-note chords* (Tymoczko 2011), in a 2-orbifold or Möbius strip. Now, we will obtain the same result by a procedure and with a notation more suitable for developing our final graphs.

There are six different 2-note chord classes, interval classes or dyads, all of them being inversionally symmetrical. They are represented in Figure 1 (left), where they are assigned interval names (m2, M2, m3, M3, P4, and Tr or tritone). The chord class 1-1 with two equal notes or unison (in fact, a multiset) is also included and represented by "X" (this uncommon notation is used instead of P1 for consistency with the next sections). The arrows show how to transform the dyads by raising one note by a semitone (or, in the opposite direction, by lowering one note by a semitone). The superscript 2 in 2–6 is its *degree of transpositional symmetry*, which doubles the arrows connecting this dyad. This diagram does not include the chord roots and represents the "local relationships" in the 2-orbifold.

Let us now represent the "global relationships" among all 2-note chords (with their roots). To do this, we group them into *voice-leading zones* (Cohn 2012, 102) or, simply, *zones* $\varphi \in [0, \ldots, 11]$. First called *sum classes* (Cohn 1998), they are the equivalence classes defined by the sum of the notes in a chord, modulo 12. For example, Bm3 = (B, D) is in the zone $\varphi = 11 + 2 = 1$ (mod 12). This way, given a chord in the zone φ, the one obtained from it by raising one note by a semitone will be in $\varphi + 1$. And chords related by *pure contrary motion*, such as FM2 = (F, G) and EM3 = (E, G♯), will be in the same zone (in this case, $\varphi = 0$).

Figure 1 (center) is a compact diagram showing the 2-note chords in the zones $\varphi = 0$ and $\varphi = 1$, while the chords in $\varphi = k$ will be those in $\varphi = k - 2$ but raising the two notes by a semitone. Note that the dyads of the same class whose roots are 6 semitones apart are in the

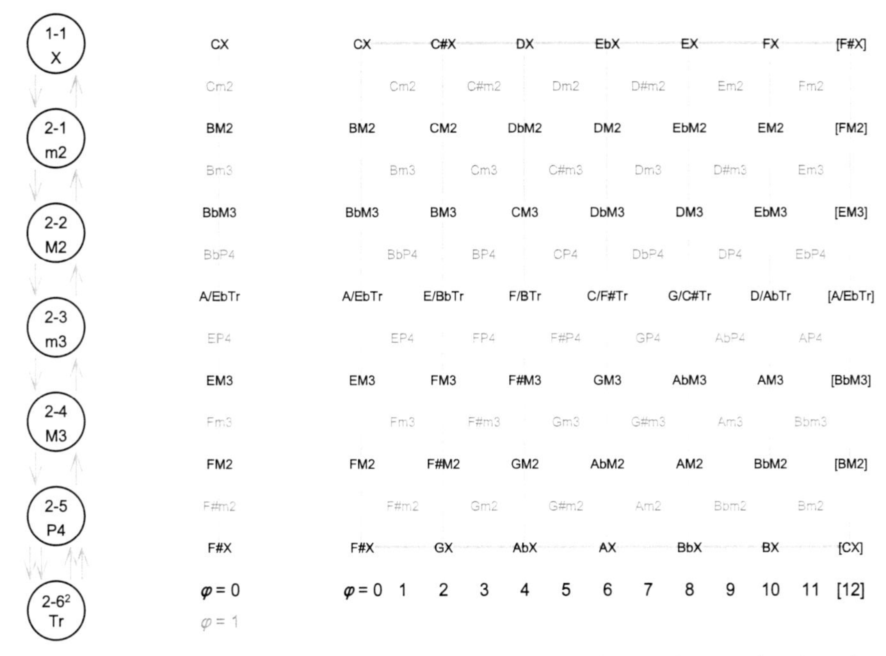

Figure 1. (Left) The 2-note chord classes plus the 2-note unison (multiset) with their single-semitonal transformations. (Center) Zones 0 and 1 of the 2-orbifold. (Right) The 2-orbifold.

same zone. And Figure 1 (right) shows all 2-note chords as given by Tymoczko (2006), but with a different notation.[2] In this diagram, each chord is transformed into the nearest ones (in oblique directions) by raising or lowering one note by a semitone, as indicated by the arrows in Figure 1 (left). Note that the tritones (2-6) are at the central horizontal axis, the perfect fourths (2-5) are one semitone apart from them, and the remaining chords are two or more semitones apart. By twisting 180° the right side of this figure and connecting it to the left one, we obtain the 2-orbifold (a Möbius strip).

3. Selection of trichords and tetrachords

There are 12 different 3-note chord classes, the trichords, 5 of them being inversionally symmetrical, while the remaining 7 can be split into two chord types related by inversion, which makes a total of 19 chord types. And there are 29 different 4-note chord classes, the tetrachords, 15 of them being inversionally symmetrical; and, by splitting the remaining 14, we obtain a total of 43 chord types. In both cases, the number of chord types is too high to obtain practical and visually simple graphs relating them. Therefore, we will just focus on the "most common" trichords and tetrachords. Let us see how to select them.

In the common practice period (around 1650–1900) the harmonies are mainly built by superimposing thirds on the 7 degrees of the major, harmonic and melodic minor scales (see, e.g., Schönberg 1983 or Piston 1988). This leads to the 4 basic triads and the 7 basic seventh chords, which are 3-10, 3-11a, 3-11b, 3-12, and 4-19a, 4-19b, 4-20, 4-26, 4-27a, 4-27b, 4-28, respectively. In addition, the augmented sixth chords add the 3-8a (Italian) and 4-25 (French). All these set types are, consequently, prevalent in Western music. On the other hand, for set classes with

[2] Tymoczko (2006) represents the 2-note chords by their actual notes, in integer notation (e.g. 48 for EM3).

Table 1. Trichord and tetrachord types considered here. A superscript on the Forte ordinal indicates the degree of transpositional symmetry, when greater than 1. An asterisk (*) means "omit 5" and a double asterisk (**) "omit ♭3". A major chord (3-11b) is normally represented by the root without any symbol. Symbol "(9)" means "add 9," whereas symbol "9" adds both the minor seventh and the major ninth. The intervallic forms start from the root.

Trichord	Symbol	Int. Form	Int.-Class Vect.
3–8a	7*	462	010101
3–8b	Ø**	642	010101
3–9	sus4	525	010020
3–10	dim	336	002001
3–11a	m	345	001110
3–11b	M	435	001110
$3–12^3$	+	444	000300

Tetrachord	Symbol	Int. Form	Int.-Class Vect.
4–19a	mΔ	3441	101310
4–19b	Δ♯5	4431	101310
4–20	Δ	4341	101220
4–21	9*	2262	030201
4–22a	(9)	2235	021120
4–22b	m4	3225	021120
4–23	7sus	5232	021030
4–24	7♯5	4422	020301
$4–25^2$	7♭5	4242	020202
4–26	m7	3432	012120
4–27a	Ø	3342	012111
4–27b	7	4332	012111
$4–28^4$	O	3333	004002

the same cardinality, the Forte ordinals are assigned so that the corresponding *interval-class vectors*[3] are arranged in decreasing lexicographic order.[4] This means that the number of smaller interval classes is progressively reduced, which arranges the set classes from the chromatic to the maximally even ones. Thus the criterion here taken is to select "full series of chord types," from the ones in the above groups having the lexicographically greatest interval-class vectors (3-8 and 4-19) to the corresponding maximally even ones (3-12 and 4-28).

Table 1 shows those trichord and tetrachord types with the symbols here used to represent them, their *intervallic forms*[5] (Nuño 2020) starting from the root, and their interval-class vectors. The added chord types are 3-8b, 3-9, 4-21, 4-22a, 4-22b, 4-23, and 4-24, which are sometimes interpreted as chromatic, incomplete or passing chords. In other musical styles, such as Pop, Latin, or Jazz, all chord types in the table are frequently used (see, for example, the list of chords given by Sher 1991, iv). Therefore, in order to keep the selected chord types to a reasonable and manageable number, as well as retaining the most relevant ones, just those in Table 1 will be considered here.

4. Parsimonious graphs

Straus (2003) gives two diagrams showing all 3- and 4-note chord classes, linked by single-semitonal transformations. Figures 2 and 3 are reduced versions of them, which only include the chord classes considered here, but splitting those being non-inversionally-symmetrical into two chord types related by inversion. These figures are analogous to Figure 1 (left), but now the arrows in opposite directions forming a pair correspond to different chord types of the same class. Similarly, multiple arrows show the different ways to move between two chord types. Among other things, splitting the two types of a set class allows us to show the relations between them, when they exist. This is the case for *P* and *L* operations between major and minor triads (Figure 2).

[3] The vector listing the number of times each of the 6 dyads is contained in a given set class or set type.

[4] Except in the case of *Z-related* pairs (two different set classes with the same interval-class vector), where one member of each pair is placed at the end of the corresponding group.

[5] The intevallic form is the sequence of intervals, in semitones, between every two adjacent pitch classes in a set type, including the interval between the last and the first ones, or any of its circular shifts.

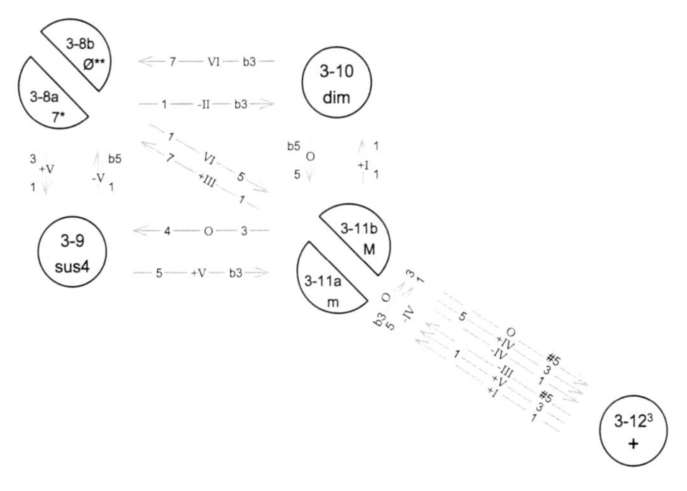

Figure 2. The 3-note chord types included in Table 1 with their single-semitonal transformations.

In these diagrams, Arabic numerals indicate the initial and final notes referring to the chord roots, where 1, 3, 4, and 5 stand for perfect or major intervals, which may be altered with ♯ or ♭, whereas major, minor, and diminished sevenths are denoted by Δ, 7, and d7, respectively; and Roman numerals at the middle of the arrows indicate the difference between the two chord roots, in semitones (letter "O" means zero). For example, Cm consists of notes (C, E♭, G) and, by raising the root (1) by a semitone, the new note is the minor seventh (7) of the new chord, a "7*" with root C + III, that is, E♭7* = (E♭, G, D♭). Or, by lowering the minor third (♭3) by a semitone, it turns into the perfect fifth (5) of the "sus4" chord with root C − V, that is, Gsus4 = (G, C, D).

This notation also allows us to easily find other parsimonious transformations, particularly $P_{0,1}$, which corresponds to two consecutive arrows where the ending note on the first matches the starting note on the second one. For example, if in Cm we raise the perfect fifth (5) by a semitone, it turns into the root (1) of a major chord; and by raising again this note by a semitone, it turns into the root (1) of a "dim" chord whose root is C − IV + I, that is, Adim = (A, C, E♭). Also, if in Cm we lower the root (1) by a semitone, it turns into the 1, 3, or ♯5 of an augmented triad; and by lowering again *the same note* by a semitone, it turns into the perfect fifth (5) of a major chord whose root is C + III, that is, E♭M = (E♭, G, B♭). This is, precisely, the *R* operation.

There are, respectively, 18 and 36 $P_{1,0}$, and 13 and 18 $P_{0,1}$ relations among the trichords and tetrachords considered here, and all but one of the $P_{0,1}$ relations can be derived with Figures 2 and 3 as explained above. However, the exception must be derived differently, as it corresponds (or may correspond) to a voice crossing between the tetrachord types "Δ♯5" and "mΔ". For example, transforming CΔ♯5 = (C, E, G♯, B), into C♯mΔ = (C♯, E, G♯, B♯) or (C♯, E, G♯, C), can be achieved by raising B by a whole tone, which crosses C. And in Figure 3, this is found by first raising the root (C) by a semitone, giving C♯m7, and then raising its minor seventh (B) by a semitone, thus avoiding the voice crossing. Cannas (2018) gives two diagrams showing both the $P_{1,0}$ and $P_{0,1}$ relations among the 4 trichord types 3-10, 3-11a, 3-11b, 3-12, and among the 5 tetrachord types 4-20, 4-26, 4-27a, 4-27b, 4-28. She also extended the analysis of tetrachords to include 4-19a, 4-19b, 4-24, and 4-25, but without providing the corresponding diagram.

After representing the local relationships among the trichords and tetrachords considered here (without indicating the roots), we will obtain the corresponding global ones (with all roots). Following the theory by Tymoczko (2011, Section 3.8), I developed Figure 4, whose left diagram is analogous to Figure 1 (center). It shows the 3-note chords in the zones $\varphi = 0$, $\varphi = 1$, and

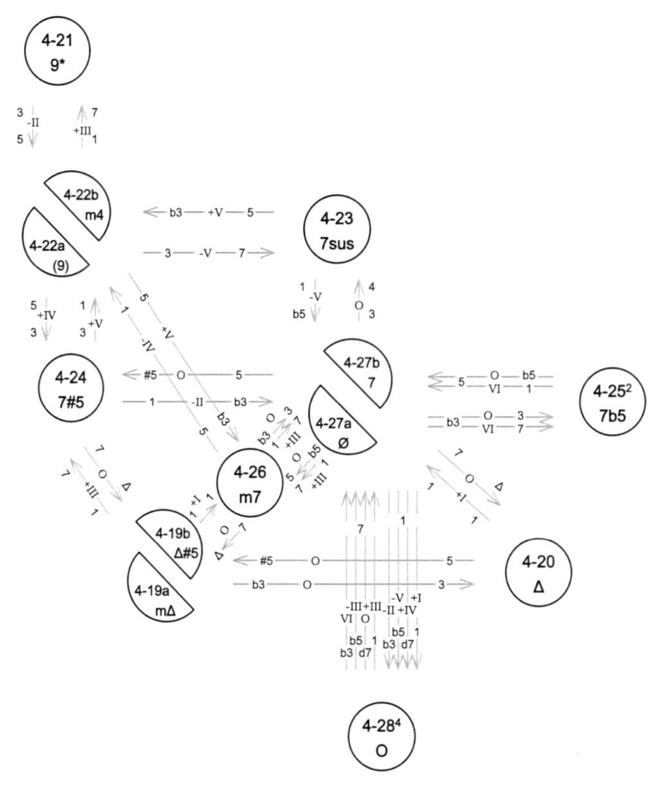

Figure 3. The 4-note chord types included in Table 1 with their single-semitonal transformations.

$\varphi = 2$, while the chords in $\varphi = k$ will be those in $\varphi = k - 3$ but raising the three notes by a semitone. Chords at the vertices are of class 1-1, but with three equal notes, and are represented by "XX." The remaining chords at the edges are dyads with one note duplicated, either the root (symbol "X") or the other note (symbol "Y"). Chords in the central regions defined by the dashed hexagons are assigned the symbols in Table 1, whereas the remaining chords are represented by the root, according to the *normal intervallic form*[6] (Nuño 2020), followed by the Forte ordinal and the letter "a" or "b" when appropriate. Note that the trichords of the same type whose roots are 4 semitones apart are in the same zone.

Superimposing all zones $\varphi \in [0, \ldots, 11]$ gives rise to a triangular prism, one of its side faces is shown in Figure 4 (right), where the oblique lines correspond to the vertical lines in Figure 1 (right) (remember that now the dyads have one note duplicated). In that prism, each chord is transformed into the nearest ones (in oblique directions with respect to the current axes) by raising or lowering one note by a semitone. The dashed hexagons also give rise to two prisms, the smaller one including the axis with the augmented triads (3-12) plus the minor (3-11a) and major (3-11b) triads, which are one semitone apart from them. And the greater hexagonal prism adds the chords being two semitones apart (chord types 3-8 to 3-10). The remaining chords are three or more semitones apart from the prism axis. Now, by twisting 120° one of the bases of the triangular prism and connecting it to the other one, we obtain the 3-orbifold, which is a "triangular Möbius strip." The result for just the trichords considered here is represented in Figure 5 in a circular graph, here called *3-Cyclops*, where φ is actually an angular position starting from "twelve o'clock" ($\varphi = 0$ for C+) and increasing clockwise. The arrows in Figure 2 are now substituted

[6] The least of all possible circular shifts of an intervallic form, with respect to the lexicographic order.

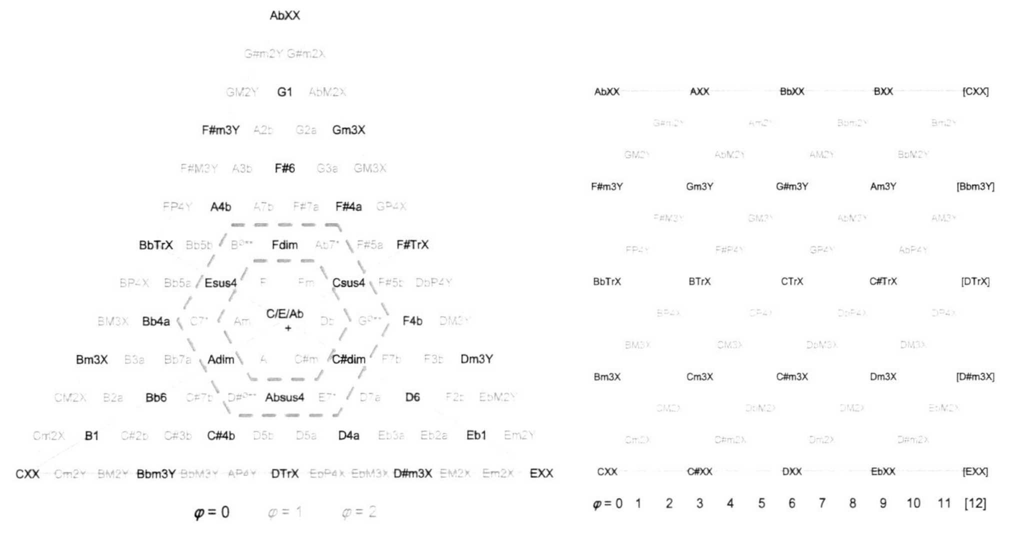

Figure 4. (Left) Zones 0, 1, and 2 of the 3-orbifold. The two central regions (dashed hexagons) contain the chord types included in Table 1. (Right) A side face of the 3-orbifold.

by lines whose directions are assumed to be clockwise and no Roman numerals are used, since the roots are directly given.

The Cube Dance by Douthett and Steinbach (1998) shows the single-semitonal transformations among the augmented, major, and minor triads, that is, those in the smaller hexagonal prism. Or, with respect to the 3-Cyclops, it just includes 1 chord type per zone. Tymoczko (2011, 105) gives an alternative representation of those chords on a cube, in the 3-orbifold. For its part, the *Tonnetz* is an earlier representation of major and minor triads, connected by *PLR* operations. On the other hand, the 3-Cyclops can be viewed as a "second-order" Cube Dance or *Tonnetz*, since it also includes the chords being 2 semitones apart from the prism axis. Thus it contains a total of 7 chord types versus 3 in the Cube Dance or 2 in the *Tonnetz*. Also, the basic operations in the *Tonnetz* are easily visualized on it: *P* and *L* are lines oblique to a circumference centered with the graph, and *R* goes through an augmented triad entering and exiting by the same letter ("a," "b," or "c"). Symbolically, $P = /$, $L = \backslash$, and $R = \wedge$. In addition, it clearly shows the *Weitzmann*[7] and *hexatonic*[8] *regions* (Cohn 2012), which correspond to zones (11,1), (2,4), (5,7), (8,10), and (1,2), (4,5), (7,8), (10,11), respectively.

A similar procedure can be carried out for the 4-note chords (Tymoczko 2011, Section 3.9), but it leads to a 4D prism whose bases are tetrahedra, which complicates the study. Additionally, the 4-note chords considered here do not correspond to a simple central region in that prism, whose axis contains the diminished seventh chords, 4-28. For example, the chords 4-21 are 4 semitones apart from them, while 4-18a and 4-18b are only 2 semitones apart and are not considered here (nor are other chords being 3 semitones apart). Therefore, only the final circular graph for the tetrachords considered here is given in Figure 6, which we will call *4-Cyclops*. Note that the tetrachords of the same type whose roots are 3 semitones apart are in the same zone.

The Power Towers of Douthett and Steinbach (1998) show the single-semitonal transformations among the diminished (4-28), half-diminished (4-27a), dominant (4-27b), and minor seventh (4-26) chords, which correspond to 1 chord type per zone in the 4-Cyclops. Cannas (2018) adds the major seventh chords (4-20), obtaining the so-called *Clover graph*. In contrast, the Douthett's *4-Cube Trio* (Cohn 2012, 158), as well as the representation by Tymoczko (2011,

[7] 3 major and 3 minor triads adjacent to the same augmented triad.

[8] 3 major and 3 minor triads lying between two consecutive augmented triads (3 zones apart).

Figure 5. The 3-Cyclops, with the 3-note chords considered in Table 1.

106) in the 4-orbifold, add the French sixth chords (4-25), which complete a 4D cube or tesseract (chord types 4-25 to 4-28). On the other hand, the 4-Cyclops can be viewed as a higher-order 4-Cube Trio, since it also includes 4-19 to 4-24. Thus, it contains a total of 13 chord types versus 5 in the 4-Cube Trio or the Clover graph, which is a high number and makes this graph more complex than the 3-Cyclops. Also, it clearly shows the *Boretz*[9] and *octatonic*[10] *regions* (Cohn 2012), which correspond to zones (1,3), (5,7), (9,11), and (11,1), (3,5), (7,9), respectively.

5. Chord patterns

The 3- and 4-Cyclops are especially suitable for representing some particular chord patterns used in musical compositions, which are given in Table 2. These patterns can also be represented on the *Tonnetz*, but only to a limited extent, since it just deals with minor (3-11a) and

[9] 4 dominant and 4 half-diminished seventh chords adjacent to the same diminished seventh chord.

[10] 4 dominant and 4 half-diminished seventh chords lying between two diminished seventh chords. These groups are 2 semitones apart, but are connected by single-semitonal transformations by means of the minor seventh or the French sixth chords.

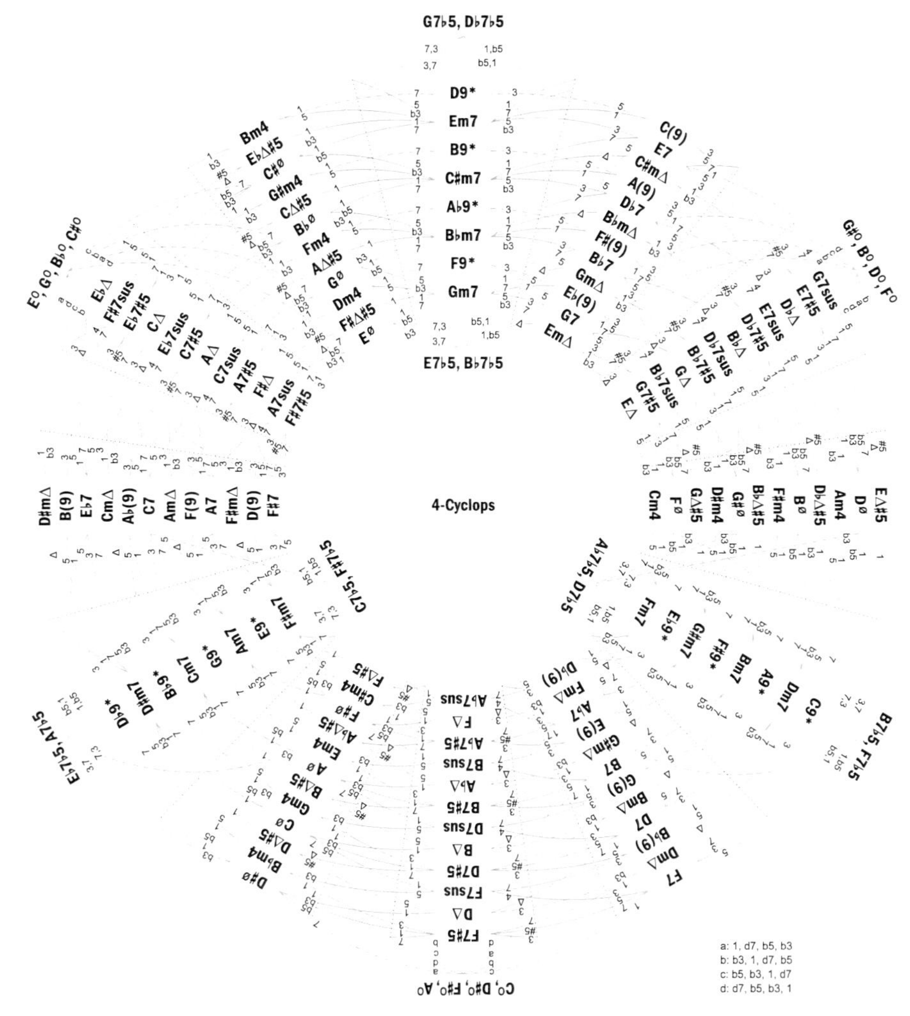

Figure 6. The 4-Cyclops, with the 4-note chords considered in Table 1.

Table 2. Particular chord patterns especially fitting the 3- and 4-Cyclops.

3-Cyclops	4-Cyclops
Parsimonious progressions of Trichords	Parsimonious progressions of Tetrachords
Same Trichord types a *major* third apart	Same Tetrachord types a *minor* third apart

major (3-11b) triads; and, when seventh chords of class 4-27 are involved, normally the "*Tonnetz* reduction" consists in omitting the seventh in the "7" chords and the root in the "Ø" chords. Cohn (2012) and Tymoczko (2011) analyze many examples of these kinds, but also including the augmented triads (3-12); and, regarding the tetrachords, they consider the five most even chord types (4-25 to 4-28). On the other hand, the 3- and 4-Cyclops include more than double the number of chord types in both cases (3-8 to 3-12 and 4-19 to 4-28, respectively), thus allowing to analyze a greater number of musical works, as well as to obtain simpler and more compact representations.

First, we will consider some examples based on trichords a major third apart, thus lying on the same zone on the 3-Cyclops, which also include parsimonious progressions. With respect to the “7” and “Ø” chords, we will use their incomplete forms, “7*” and “Ø**”, which are better approximations to the real chords than those used with the *Tonnetz* and, what is very advantageous, they lead to more compact representations.

Let us start with Beethoven’s Sonata for Violin and Piano in F major, Op. 24. The harmonies in the 2nd mvt., mm. 38–54, are the following:

$$\{B\flat m\ \ \%\ \}\ \{G\flat\ \ D\flat 7\ \ \%\ \ G\flat{-}C\flat\ \ G\flat{-}D\flat 7\ \ G\flat\}\ \{F\sharp m\}$$

$$\{D{-}G\ \ D{-}A7\ \ D\}\ \{Dm\}\ \{F7\ \ B\flat{-}E\flat\ \ B\flat{-}F7\ \ B\flat\}$$

where each chord or each pair linked by a dash lasts one measure and symbol “%” means to repeat the previous measure. Chords related to the same consonant triad are grouped in curly brackets. This chord progression is represented in Figure 7 on the 3-Cyclops, where the initial chord is specially marked. The 3 minor chords (B♭m, F♯m, Dm) are a major third apart in descending order, as are the 3 major chords related to them by *L* and *P* operations (G♭, D, B♭). The latter are affirmed by cadences including the dominant seventh and subdominant chords, each group lying in one zone. Since we used the incomplete form of the “7” chords, the result is very compact, only requiring 3 nearby zones: 4, 5, and 8. If we had used the “7” chords with the seventh omitted, as is usual with the *Tonnetz*, they would have lain in the zone 2 of Figure 5. And regarding their 4-note forms, they lie in different zones (1, 5, 9) of Figure 6 and are not grouped together.

We will now analyze the mm. 23–43 of the Consolation in D♭ major, Op. 102, No. 3 by Liszt, whose harmonies are

$$\{D\flat\}\ \{G^{\varnothing}\ \ G^{\varnothing}{-}C7\ \ Fm\ \ \%\ \ C7/F\ \ Fm\}\ \{C7/F\ \ F\ \ \%\ \}$$

$$\{Am\ \ Am{-}E7\ \ Am\ \ E7\ \ Am\}\ \{E7/A\ \ A\ \ \%\ \}\ \{D\flat\ \ A\flat 7\ \ D\flat\}$$

where some chords are played over a pedal note, here represented by a slash followed by the pedal. This chord progression is represented in Figure 8 on the 3-Cyclops (without the pedals) and can be compared with Cohn (2012, 187), who also provides a Web animation. Now the 3 major chords (D♭, F, A) are a major third apart but in ascending order, and there are only 2 minor chords (Fm, Am), related to them by *L* and *P* operations, which are affirmed by longer cadences. A “Ø” chord is now included, whose incomplete form, together with those of the “7” chords, make the representation really compact, just requiring 2 consecutive zones (1 and 2). In fact, the 3-Cyclops is especially suitable for representing the cadences V7–I(m) and $II^{\varnothing}$–V7–I(m) with major or minor tonic chords. The jazz tune “Giant Steps” by Coltrane (Sher 1991) is closely related to this, as it just consists of cadences V7–IΔ and IIm7–V7–IΔ a major third apart.

Regarding examples with the 4-Cyclops, let us start with the Piano Concerto No. 2 in C minor, Op. 18, by Rachmaninoff. In the 1st mvt., mm. 1–8, there is a pure single-semitonal progression,

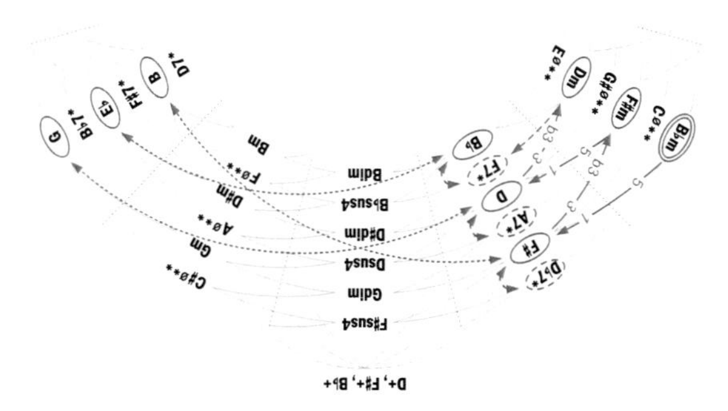

Figure 7. Beethoven, Sonata for Violin and Piano in F major, Op. 24, 2nd mvt., mm. 38–54.

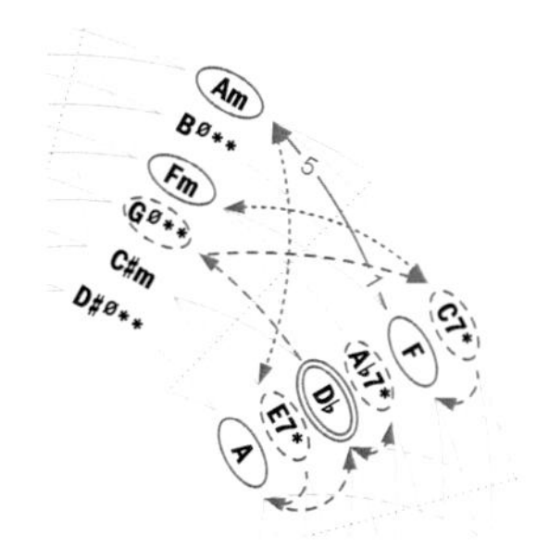

Figure 8. Liszt, Consolation No. 3, Op. 102, mm. 23–43.

represented in Figure 9 on the 4-Cyclops with a simple line:

$$[\mathrm{Fm}(5)]\ \ \mathrm{D}\flat\Delta\ \ \mathrm{D}^{\varnothing}\ \ \mathrm{Fm7}\ \ \mathrm{F7}\ \ \mathrm{Fm7}\ \ \mathrm{D}^{\varnothing}\ \ \mathrm{D}\flat\Delta$$

Here, a note in parentheses means to add that note to the chord. Thus Fm(5) is Fm with two Cs. This chord is written in brackets because it does not appear in the 4-Cyclops, but was included in the figure to illustrate the example. They are precisely the two Cs that move up or down by semitone throughout the progression, except going to F7. A pedal F–C (in three octaves), which belongs to all the harmonies, gives consistency to the full chord progression. There is another pedal A♭ (in two octaves), except in F7. The first chord moves to D$^{\varnothing}$ through D♭Δ instead of D^{O}, possibly because the latter does not contain the pedal C and, additionally, it has two tritones and the former none.

Our next example is *Indudable* (Bossa Nova) by Nuño (2012), whose mm. 19–27 consist of the following chords (actually, some of them include additional tensions):

G♯m7 C♯Δ Fm7 B♭Δ Dm7 G6 Bm7 E7sus G♯m7

This chord progression is represented in Figure 10 on the 4-Cyclops. The 4 minor seventh chords (G♯m7, Fm7, Dm7, Bm7) are a minor third apart, thus lying on the same zone. With respect to the other chords, their roots are also a minor third apart, but instead of having the uniform sequence C♯Δ, B♭Δ, GΔ, EΔ, the last two chords (marked with dashed lines in Figure 10) are replaced by G6 (enharmonic to Em7) and E7sus, respectively. Even so, the representation is again simple and compact.

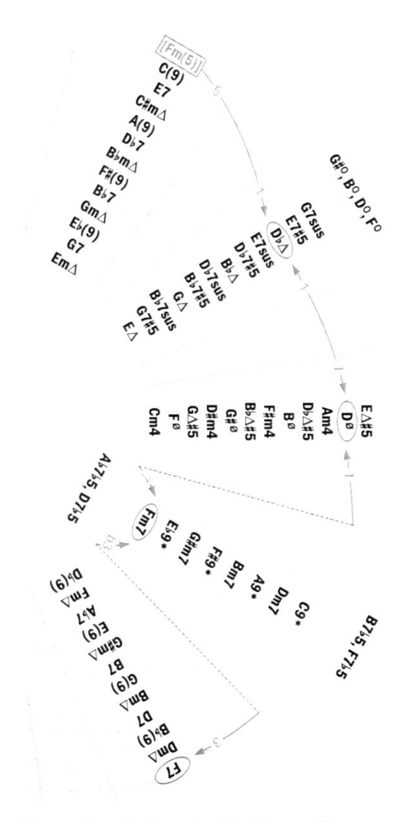

Figure 9. Rachmaninoff, Piano Concerto No. 2, Op. 18, 1st mvt., mm. 1–8.

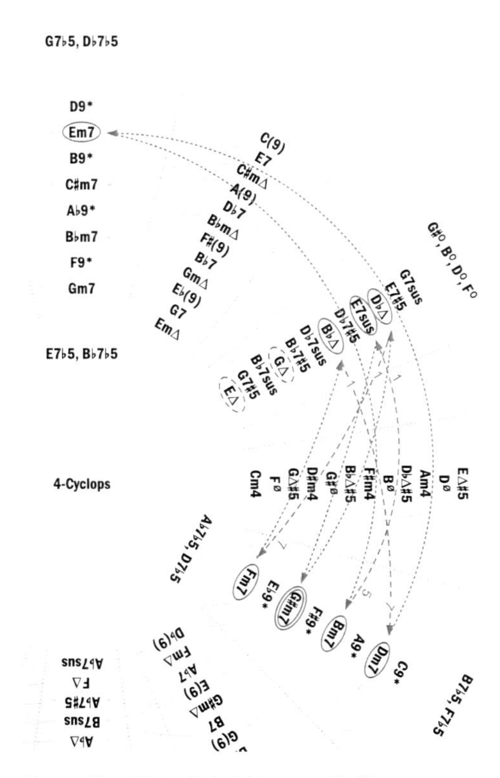

Figure 10. Nuño, *Indudable*, mm. 19–27.

Figure 11. Chopin, Prelude in E minor, Op. 28, No. 4, mm. 1–12. Melody and harmonic structure.

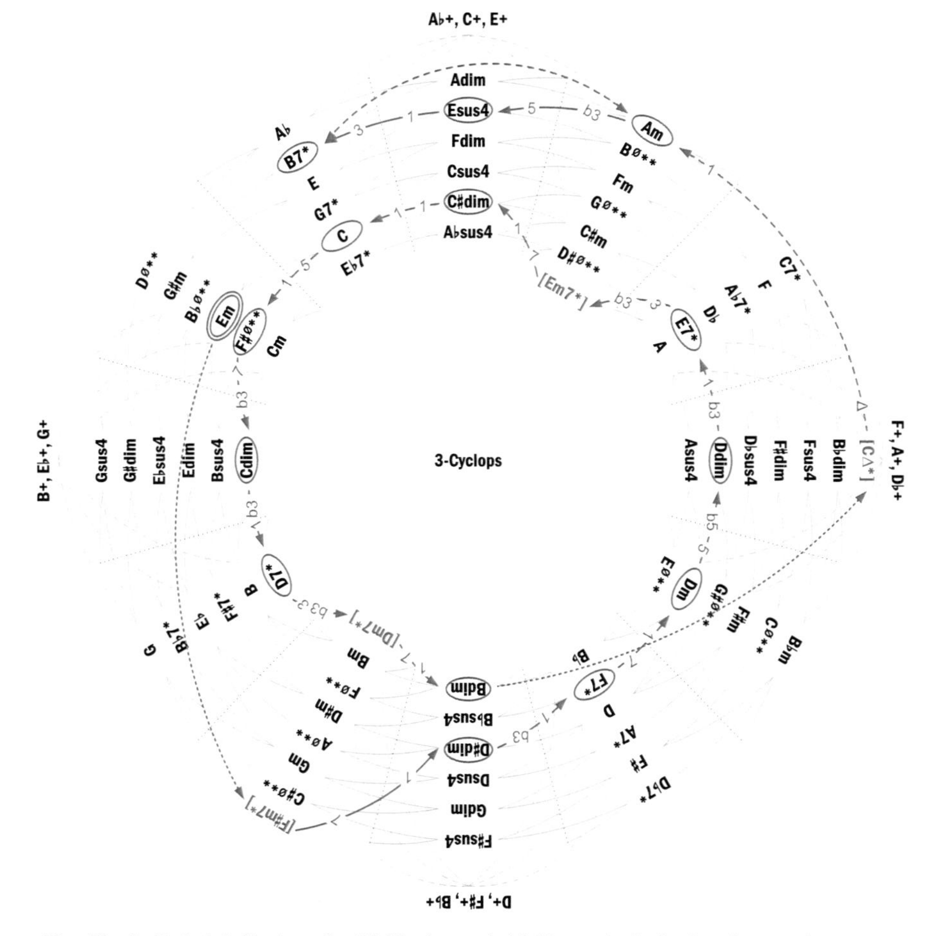

Figure 12. Chopin, Prelude in E minor, Op. 28, No. 4, mm. 1–12. Harmonies in the three lower voices.

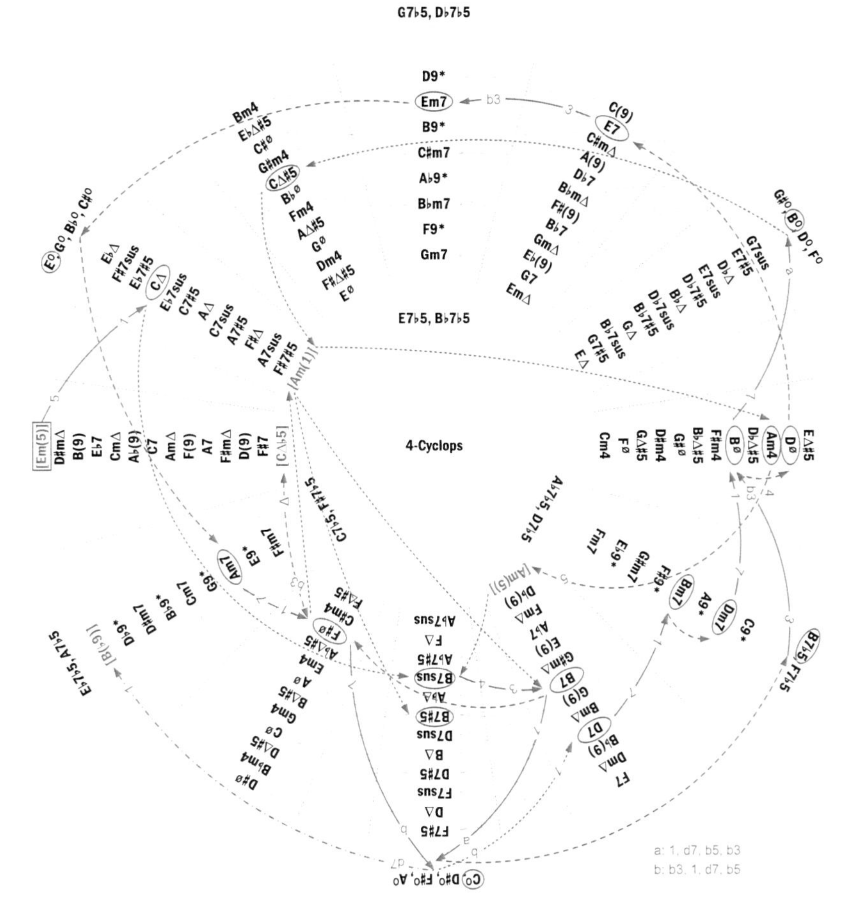

Figure 13. Chopin, Prelude in E minor, Op. 28, No. 4, mm. 1–12. Full harmonies.

The last example is Chopin's Prelude in E minor, Op. 28, No. 4, one of the most interesting pieces analyzed by Tymoczko (2011, 287–293) and Cohn (2012, 160–166), the latter providing a Web animation. Figure 11 is a simplified score with mm. 1–12. As will be seen, this composition is best understood by first analyzing the harmonies in the three lower voices, represented in Figure 12 on the 3-Cyclops. They pass through all the trichord types considered in this graph, except the augmented triads (perhaps too dissonant?). Chopin also included the chord types "m7*" (3-7a) and "Δ*" (3-4a), defined by the intervallic forms {372} and {471}, which are the incomplete tonic seventh chords in natural minor and major keys, respectively. From the second chord (F♯m7*), the three lower voices strictly follow a descending single-semitonal ($P_{1,0}$) line, covering more than one full turn on the graph. Then, other parsimonious transformations are employed to finish the phrase, as indicated in the score.

On the other hand, the austere melody also draws a descending line, B–A–G♯–F♯, which completes the harmonies and leads to a more complex representation on the 4-Cyclops (Figure 13). Apart from the chord types considered in this graph, Chopin also included "(♭9)" (4-18a) and "Δ♭5" (4-16a), defined by {1335} and {4251}, respectively.

6. Conclusion

Two novel graphs, called Cyclopes, are presented, which relate the most common trichords and tetrachords by single-semitonal transformations. They include more than double the number of chord types in previously published graphs, thus allowing one to analyze a larger repertoire in a practical way. They are especially suitable for representing parsimonious chord progressions, trichords a major third apart, tetrachords a minor third apart, and the cadences V7–I(m) and $II^{\emptyset}$–V7–I(m) with major or minor tonic chords. In all those cases, the results are simple and compact, thus allowing one to clearly visualize the relationship among the corresponding chords and better understand those composition patterns, as well as being efficient mnemonic resources. Consequently, they prove to be practical tools that can be used both for music analysis and composition.

Acknowledgments

I thank editor Darrell Conklin and the anonymous reviewers for their valuable comments, which contributed to improving the quality of this paper.

Disclosure statement

No potential conflict of interest was reported by the author.

Supplemental data

Supplemental data for this article can be accessed online at https://doi.org/10.1080/17459737.2021.1923844.

ORCID

Luis Nuño http://orcid.org/0000-0001-8486-0582

References

Callender, Clifton, Ian Quinn, and Dmitri Tymoczko. 2008. "Generalized Voice-Leading Spaces." *Science* 320 (5874): 346–348.

Cannas, Sonia. 2018. "Geometric Representation and Algebraic Formalization of Musical Structures." Ph.D. dissertation, Université de Strasbourg and Università degli Studi di Pavia e di Milano-Bicocca.

Cohn, Richard. 1996. "Maximally Smooth Cycles, Hexatonic Systems, and the Analysis of Late-Romantic Triadic Progressions." *Music Analysis* 15 (1): 9–40.

Cohn, Richard. 1997. "Neo-Riemannian Operations, Parsimonious Trichords, and Their 'Tonnetz' Representations." *Journal of Music Theory* 41 (1): 1–66.

Cohn, Richard. 1998. "Square Dances with Cubes." *Journal of Music Theory* 42 (2): 283–296.

Cohn, Richard. 2012. *Audacious Euphony: Chromatic Harmony and the Triad's Second Nature*. New York: Oxford University Press.

Douthett, Jack, and Peter Steinbach. 1998. "Parsimonious Graphs: A Study in Parsimony, Contextual Transformations, and Modes of Limited Transposition." *Journal of Music Theory* 42 (2): 241–263.

Forte, Allen. 1973. *The Structure of Atonal Music*. New Haven: Yale University Press.

Gollin, Edward. 2005. "Neo-Riemannian Theory." *Zeitschrift der Gesellschaft für Musiktheorie (ZGMTH)* 2 (2–3): 153–155.

Nuño, Luis. 2012. *Puesta de Sol*. Vol. 1. Madrid: Acordes Concert, S.L.

Nuño, Luis. 2020. "A Detailed List and a Periodic Table of Set Classes." *Journal of Mathematics and Music* 1–21. https://doi.org/10.1080/17459737.2020.1775902.

Piston, Walter. 1988. *Harmony*. 5th ed. New York: W. W. Norton and Co.

Schönberg, Arnold. 1983. *Theory of Harmony*. 3rd ed. Berkeley, Calif.: University of California Press.

Sher, Chuck. 1991. *The New Real Book*. Vol. 2. Petaluma, Calif.: Sher Music Co.

Straus, Joseph. 2003. "Uniformity, Balance, and Smoothness in Atonal Voice Leading." *Music Theory Spectrum* 25 (2): 305–352.

Tymoczko, Dmitri. 2006. "The Geometry of Musical Chords." *Science* 313 (5783): 72–74.
Tymoczko, Dmitri. 2011. *A Geometry of Music: Harmony and Counterpoint in the Extended Common Practice*. New York: Oxford University Press.
Waller, Derek A. 1978. "Some Combinatorial Aspects of the Musical Chords." *The Mathematical Gazette* 62 (419): 12–15.

Triadic patterns across classical and popular music corpora: stylistic conventions, or characteristic idioms?

David R. W. Sears and David Forrest

Many musical traditions – from Western art, to popular and commercial – organize pitch phenomena around a referential pitch class (or *tonic*) and feature triads and seventh chords. As a result, triadic progressions associated with one tradition sometimes resurface in others. How, then, are we to distinguish between the *conventional* harmonic patterns that span several time periods, and the *characteristic* idioms that delimit a single period?

This essay presents a comparative study of triadic progressions in four data sets comprised of expert harmonic annotations: Annotated Beethoven Corpus (ABC), Theme and Variation Encodings with Roman Numerals (TAVERN), Rolling Stone-200 (RS-200), and McGill Billboard (Billboard). Using methods for counting, filtering, and ranking multichord expressions, we reveal conventional and characteristic progressions and examine broad trends over time. We also include an accompanying standalone application that allows users to adjust various stages of the model pipeline and export the data for further exploration and analysis.

1. Introduction

Contemporary scholarship often restricts definitions of *tonality* to the major-minor diatonic system found in European music from about 1600 to around 1910, the period of "common practice" (Hyer 2002). Defined in this way, tonal music thereby excludes styles that nevertheless share a preference for pitch centricity and tertian pitch structures. Such is the case with many of the vernacular musical traditions from the twentieth and twenty-first centuries, which employ a variety of diatonic systems (modal, major-minor, pentatonic, etc.), but still organize pitch phenomena around a referential pitch class (or *tonic*), and regularly feature triads, seventh chords, and various (tertian) extensions. As a result, many of the triadic patterns associated with common-practice music resurface in later musical traditions.

How, then, are we to distinguish between the conventional harmonic utterances that span several historical time periods, and the characteristic idioms that delimit a single period? If, indeed, a *style* is nothing more than a storehouse of the replicated patterns found therein (Meyer 1989), then knowledge of the style will depend on a listener's understanding of its recurrent *and* distinguishing features. From a statistical point of view, patterns are thus *conventional* to one or more styles if they recur frequently, but *characteristic* of one or more styles if they fail to appear in others (Huron 2001). Of course, the two concepts are not mutually exclusive – a pattern can be both conventional to, and characteristic of, a given style – but neither are they synonymous: the most conventional patterns are not necessarily the most characteristic, and vice versa.

By way of example, the simple and compound cadences in Figure 1(a) are perhaps the most replicated, and thus, conventional voice-leading formulæ in the tonal system (Sanguinetti 2012). By comparison, many of the patterns described in Fenaroli's rule of the octave, shown in Figure 1(b), are restricted to – and thus, characteristic of – the instrumental repertories associated with the long eighteenth century (Gjerdingen 2007). Similarly, the double-plagal, aeolian, and blues progressions in Figure 1(c) serve as characteristic style markers for rock music (Biamonte 2010), whereas the doo-wop and axis progressions in Figure 1(d), which observe the functional temporal relations associated with common-practice tonal harmony, may be conventional in both traditions.[1] Providing evidence to support these claims, however, would require a comparative study that determines the prevalence and distinctiveness of these patterns in various classical and popular music corpora.

Thus, the goal of this study is to identify and compare the conventional and characteristic harmonic patterns associated with (1) common-practice European music; and (2) the Anglophone popular music traditions of the latter half of the twentieth century. To that end, we have selected four data sets comprised of expert annotations of tertian harmony and adapted the analysis pipeline developed in Sears and Widmer (2020) for the discovery of recurrent three- and four-chord progressions. We begin in Section 2 by presenting the selected data sets and describing the encoding scheme that converts the various chord symbols into a single unifying chord typology. Section 3 then describes the methods for identifying, counting, filtering, and ranking harmonic progressions. Section 4 presents the most conventional and characteristic harmonic progressions in the selected corpora and examines broad trends associated with the prevalence of these patterns over time. Finally, Section 5 considers limitations and directions for future research and introduces the *Triadic Harmony Analysis Tool*, an accompanying standalone application that allows users to examine recurrent patterns in the selected corpora.

2. Triadic harmony: the data

For our purposes, "triadic harmony" will refer to the tertian pitch structures found in multi-voiced music where pitch centricity is a fundamental organizing principle, regardless of style or time period. This definitional inclusivity not only embraces genres from many musical traditions – renaissance madrigals, classical sonatas, rock ballads, and so on – but also allows us to identify the triadic patterns that are both conventional within and across styles, and characteristic of a specific style. How, then, might we identify the triadic patterns associated with the common-practice and popular music traditions?

Previous studies have identified multi-voiced patterns using 'bottom-up' representation schemes that start from the written or recorded trace (e.g. a notated score or audio recording), or "top-down," theoretically motivated schemes that start from harmonic transcriptions of the musical surface (White 2015). Although top-down schemes vary with respect to the selected typology (Roman numerals, figured bass symbols, pop-chord symbols, etc.), and often depend on expert human annotators (e.g. Burgoyne, Wild, and Fujinaga 2011; Declercq and Temperley 2011; Tymoczko 2011), corpus studies often prefer top-down schemes for musical traditions that do not rely on a consistent visual notation system. Indeed, the recent surge of interest in developing hand-curated data sets of harmonic annotations has sparked a flurry of computational studies examining various genres and style periods, though typically in isolation (e.g. Conklin 2010;

[1] Richards (2017) points out, however, that the axis progression may imply several tonal centers simultaneously.

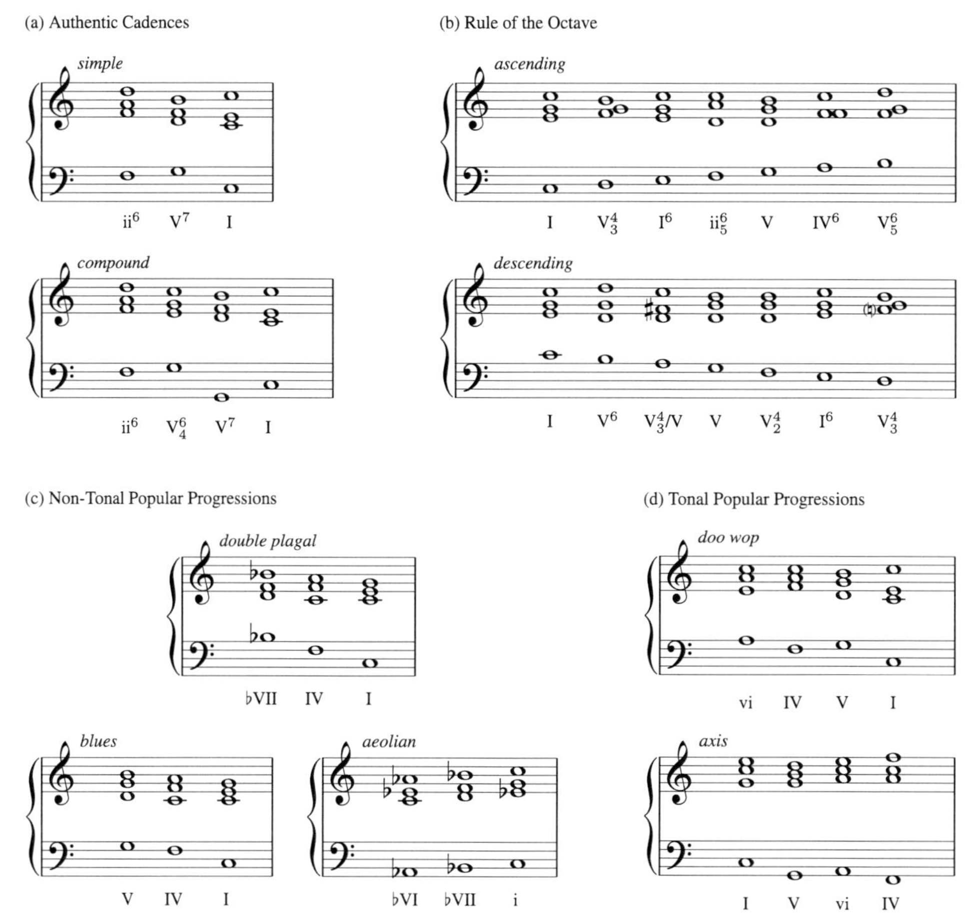

Figure 1. Common triadic patterns in the common-practice and popular music traditions. (a) Common-practice authentic cadential progressions that allot one (*simple*) or two (*compound*) metrical units to the dominant. (b) Fenaroli's Rule of the Octave in first position, ascending and descending. (c) Popular music non-tonal progressions. (c) Popular music tonal progressions.

Shanahan and Broze 2012; Moss et al. 2019). Thus, the time is ripe for a comparative study of triadic patterns across both classical and popular music traditions.

Table 1 provides descriptive statistics for the data sets selected for this study. The Annotated Beethoven Corpus (ABC) consists of harmonic analyses of all Beethoven string quartets encoded in MusicXML (Neuwirth et al. 2018).[2] Each movement includes Roman numeral annotations encoded as a regular expression that incorporates the key symbol, Roman numeral, chord form, figured bass symbol, and extensions and suspensions. The Theme and Variation Encodings with Roman Numerals (TAVERN) data set consists of 27 theme and variation sets by Mozart and Beethoven, with the notated movements and harmonic annotations encoded in standard kern format (Devaney et al. 2015).[3] Like the ABC data set, TAVERN also employs a Roman numeral chord typology, but also provides annotations for the harmonic functions within each phrase (e.g. tonic, predominant, dominant). Together, these data sets include over 40,000 chords and will represent the *classical* corpus in the analyses that follow.

[2] https://github.com/DCMLab/ABC.
[3] http://getTAVERN.org.

Table 1. Data sets and descriptive statistics for the corpus.

Data set	N_{pieces}	N_{chords}	Years	Style	Boundaries/Piece		
					N_b	$\%_b$	IC_b
ABC	70	27,993	1801–1826	classical	61 (28)	15 (02)	1.93 (1.78)
TAVERN	27	12,444	1765–1810	classical	75 (33)	16 (01)	1.09 (1.20)
RS-200	200	19,433	1949–2002	popular	19 (18)	19 (08)	1.06 (1.44)
McGill Billboard	721	95,522	1958–1991	popular	22 (13)	16 (05)	1.78 (2.32)

Note. N_{pieces} includes pieces that are not duplicated in the other data sets and consist of at least one chord annotation. N_{chords} does not include symbols associated with repeated or nonchordal events. Estimates in Boundaries/Piece reflect the mean and standard deviation for each data set. N_b denotes the mean number of boundary events in each piece. $\%_b$ denotes the mean percentage of boundary events relative to the number of events in each piece. IC_b denotes the mean information content estimated by IDyOM for the boundary events in each piece.

Of the many published data sets representing Western popular music, two data sets were selected for this study that include both key and chord annotations, and sample widely (i.e. across artists) from a given corpus. The Rolling Stone Corpus (RS-200) consists of Roman numeral annotations for 200 songs sampled from Rolling Stone magazine's "500 greatest songs of all time" (Declercq and Temperley 2011).[4] The McGill Billboard data set consists of expert annotations for 890 songs sampled from the Billboard "Hot 100," of which 721 represent unique songs that were not shared with the RS-200 data set and contain at least one chord annotation (Burgoyne, Wild, and Fujinaga 2011).[5] Unlike the RS-200 data set, however, the Billboard data set selects a roughly equal number of songs from each decade and employs a pop-chord slash-notation scheme (Harte et al. 2005), which encodes the root, quality, and inversion symbols for primarily tertian sonorities, including triads, seventh chords, and their various extensions. Together, these data sets include over 100,000 chords and will represent the *popular* corpus.

For each piece, symbols indicating chord repetitions and nonchordal events were removed. The remaining symbols in each data set represent various properties of the chords within a specified tonal context, including their roots, qualities, inversions (or bass notes), extensions, and nonchord tones. Due to differences in the encoding schemes across data sets, however, we converted the chord symbols from each data set into a single, standard representation scheme. The annotation format consists of four parts.

$$\langle \texttt{Roman Numeral} \rangle \langle \texttt{Quality} \rangle \langle \texttt{Inversion} \rangle \langle \texttt{Extensions} \rangle$$

Given the diversity of diatonic systems reflected in these data sets, chord roots were encoded using Roman numerals in relation to the prevailing key. Following Biamonte (2010, 97), Roman numerals derived from the major scale were treated as normative, but with the expressed understanding that the presence of chromatic alterations (e.g. ♭, ♯, etc.) does not presume chromaticism. A harmony like ♭VII, for example, is likely to be diatonic in many modal and pentatonic systems, if not in the major-minor system. Finally, the relative-root encoding scheme represents presumably chromatic harmonies like applied (or secondary) dominants and augmented-sixth chords according to the prevailing key, in some cases using chromatic alterations (e.g. vii/V→ `♯iv`). Thus, a major triad built on the second scale step in the minor pentatonic mode would receive a flat symbol on the third major-mode scale-degree (`♭III`).

Following convention, the case of the Roman numeral was used to denote the major or minor quality of the triad (upper case indicates major), with further symbols included to represent other triad and seventh-chord qualities (e.g. augmented triad→`+`; half-diminished→`h`; power→`p`, etc.). To represent inversions, bass notes representing members of the constituent triad or seventh

[4] http://rockcorpus.midside.com/.

[5] https://ddmal.music.mcgill.ca/research/The_McGill_Billboard_Project_(Chord_Analysis_Dataset)/.

chord were represented using figured bass symbols (e.g. first inversion triad→ `6`; second inversion seventh→ `43`, etc.). Finally, tertian extensions, suspensions and other embellishing tones (including those in the bass voice), and added tones were encoded in parentheses, with added tones preceded by a plus symbol (e.g. V9→ `V(9)` sus4→ `(4)`; added sixth→ `(+6)`, etc.). Following Neuwirth et al. (2018), cadential six-four harmonies were encoded as double suspensions above the dominant (`V(64)`). The parsers for each data set, which include a complete description of the encoding scheme, are available for download.[6]

3. Pattern discovery: the methods

In many ways, digitized symbolic expressions of harmonic progressions are not unlike the lexicalized recurrent word combinations found in natural language corpora. An expression like *stiff drink*, for example, is represented as a unidimensional sub-string extracted from a longer string (e.g. *I need a stiff drink*), and features events that appear at or near the surface (i.e. *stiff* and *drink* appear contiguously). Corpus linguists generally call these combinations *collocations*, but the term admits multiple meanings in language research, and so has recently been superseded by the less ambiguous term *multiword expression* (Evert 2008). In our case, a *multichord expression* (MCE) can therefore serve an analogous function in music research as a recurrent and predictable chord combination that is directly observable in a corpus.

To identify MCEs in the data sets selected for this study, we might start by dividing a given sequence of chord labels into contiguous sub-sequences of cardinality n, called n-grams, and then counting the instances (or *tokens*) associated with each distinct n-gram *type*, denoted by $\mathcal{T}$. This process produces an n-gram list, which consists of a list of the distinct n-gram types ordered by a ranking function of some sort.

Following an initial count of n-gram types, collocation discovery algorithms employ filtering methods that remove irrelevant n-gram types and/or tokens from the list. Perhaps the most obvious approach is to exclude types that contain too few tokens to justify closer examination, though the selection of a frequency threshold is often somewhat arbitrary (Sears and Widmer 2020). Given the motivations for the present study, we might also exclude n-gram tokens whose constituent members cross a phrase boundary. Grouping and boundary perception are deemed to be fundamental processes in several computational tasks, such as speech segmentation (Brent 1999), word discovery (Jusczyk 1997), and melodic and harmonic prediction (Sears et al. 2018). In the context of MCEs, we could therefore assume that a given token is far less likely to serve as a meaningful utterance if its terminal member does not precede a segment boundary.

This can be particularly problematic for harmonic loops in popular music, which elide the initial and terminal functions of a repeating progression. Biamonte (2010) has argued, for example, that the aeolian progression is *closed* if tonic harmony serves as its terminal member, ♭VI-♭VII-i, but a looped aeolian progression will necessarily include *open* rotations like ♭VII-i-♭VI and i-♭VI-♭VII.[7] Including each of these rotational (or *cyclic*) patterns results in combinatorial redundancy in the n-gram list (Lartillot 2005). Thus, an inventory of harmonic patterns in these corpora might exclude open rotations by considering the grouping principles that lead listeners to segment the musical surface.

Although each of the corpora examined here includes annotations for segment boundaries of one sort or another, these annotations differ with respect to the level(s) of the grouping hierarchy they represent (i.e. motive, phrase, theme, etc.), and so are more prevalent in some data sets than others. Thus, following Pearce, Müllensiefen, and Wiggins (2010), we determined event

---

[6] https://osf.io/kdzm3/.

[7] Biamonte (2010) also includes neighboring, passing, and circular functions among the possible progression types.

boundaries based on a probabilistic implementation of Event Segmentation Theory (EST), which argues that discontinuities over time elicit prediction errors that force the perceptual system to segment activity into discrete time spans (Kurby and Zacks 2008). These discontinuities can take many forms, but we focus on the harmonic sequences themselves, using the Information Dynamics of Music (IDyOM) model to generate a probabilistic prediction for each chord in the sequence (Pearce 2005). Given EST, IDyOM therefore assumes that segment boundaries result from prediction errors, either because the context fails to stimulate strong expectations for any continuation, or because the actual continuation is unexpected (Meyer 1957; Narmour 1990). In this respect, IDyOM reflects the statistical approaches to event segmentation found in natural language contexts (Saffran 2003), wherein listeners acquire knowledge of the underlying vocabulary of a natural language via a domain-general statistical learning mechanism.

The model is presented in detail elsewhere (Pearce 2005), but in short, IDyOM estimates the conditional probability of event e_i in the sequence given a preceding context of $n-1$ events, or e_{i-n+1} to e_{i-1}, notated here as e_{i-n+1}^{i-1}. Given a model estimate for $p(e_i \mid e_{i-n+1}^{i-1})$, the unexpectedness of e_i can be defined as the *information content* (*IC*).

$$IC(e_i \mid e_{i-n+1}^{i-1}) = \log_2 \frac{1}{p(e_i \mid e_{i-n+1}^{i-1})}. \tag{1}$$

IC therefore represents the degree of contextual unexpectedness or surprisal associated with e_i.

Using the IC estimates produced by IDyOM, Pearce, Müllensiefen, and Wiggins (2010) identified a boundary if $IC(e_i)$ was higher than the immediately surrounding events, $IC(e_{i-1})$ and $IC(e_{i+1})$, and then implemented a continuous measure of boundary strength for e_i (Brent 1999). In our view, however, this method is more likely to identify *initiating* rather than *closing* events. Shown in Figure 2(a), a measure of boundary strength should privilege patterns featuring highly predictable events at $IC(e_{i-1})$ and $IC(e_i)$ (e.g. V-I), followed by a prediction error at $IC(e_{i+1})$ (e.g. iii^6). Thus, we estimate the boundary strength b for e_i if $IC(e_i)$ is lower than the immediately surrounding events, $IC(e_{i-1})$ and $IC(e_{i+1})$, using the following measure.

$$b(e_i) = \begin{cases} \dfrac{IC(e_{i+1}) - IC(e_i)}{(IC(e_{i-1}) - IC(e_i)) \times IC(e_i)}, & \text{if } IC(e_i) < IC(e_{i-1}) \wedge IC(e_i) < IC(e_{i+1}), \\ 0, & \text{otherwise.} \end{cases} \tag{2}$$

Following Pearce, Müllensiefen, and Wiggins (2010), we included a token in the final n-gram list if the boundary strength of its terminal member e_i was greater than a threshold of k standard deviations above an exponentially weighted moving average with a window of 20 events, where $k = 1$.[8] Shown in Table 1, approximately 15–19% of the events in each data set met these criteria. Figure 2(b) presents an example bar plot of the boundary strength estimates for the annotations from the initial theme in Beethoven's WoO 63 in the TAVERN data set. The dotted line represents the weighted moving average, and each boundary event is shown with an open circle marker.

With a filtering method in place, pattern discovery algorithms finally include a ranking function that sorts the final list of n-gram types not by their counts, but by some measure of probabilistic inference, either in reference to a null model (Conklin and Bergeron 2008; Collins et al. 2016; Sears and Widmer 2020), or to a comparison corpus (or *anticorpus*) (Conklin 2010; Conklin and Bergeron 2010). Statistical *association* (or *attraction*) measures, for example, assume that an expression is salient, important, or memorable to the degree that

[8] Pearce, Müllensiefen, and Wiggins (2010) employ a linearly weighted moving average, with a window starting from the first event in the piece. Here, we employ an exponential function in order to simulate the effects of memory decay, and also select a moving window of 20 events to ensure a uniform distribution of boundary events across the piece.

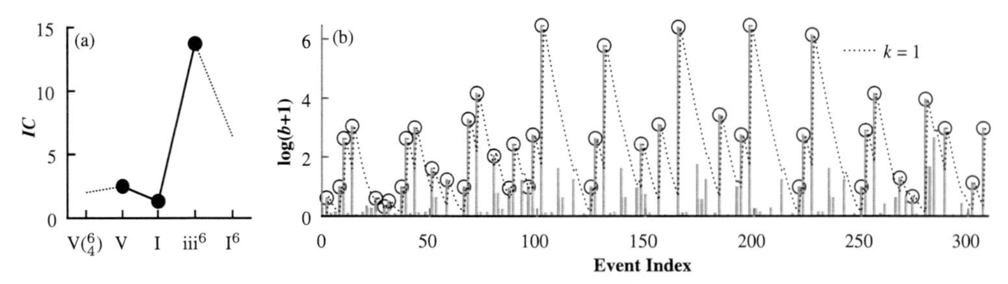

Figure 2. (a) Line plot of IC estimates for a cadential progression, V($^6_4$)-V-I, followed by an initiating progression starting with iii^6-I^6. The measure of boundary strength expressed in Equation (2) privileges patterns like this one (shown with closed circles) that feature highly predictable events at $IC(e_{i-1})$ and $IC(e_i)$, followed by a prediction error at $IC(e_{i+1})$. (b) Example bar plot of the boundary strength estimates b for the sequence of chord annotations from the theme in Beethoven's WoO 63 in the TAVERN data set. The dotted line represents the 1 SD threshold above a weighted moving average with a window of 20 events, with open circles denoting boundary events (i.e. events above the threshold).

its observed frequency in a corpus is greater than a hypothesized expected frequency (or *background probability*), often operationalized as the joint probability of the constituent members (Evert 2008).

The number of association measures is admittedly large (Pecina 2005), but Sears and Widmer (2020) compared four of the most well-known measures in a discovery task for voice-leading patterns and found that a weighted measure of *pointwise mutual information* (*pMI*) outperformed the other measures, so we adopted that method here.

The *pMI* measure estimates the ratio of the observed probability of an n-gram type $\mathcal{T}$ consisting of events e_1^n from the final n-gram list (i.e. after filtering), to the joint probability of its constituent members.

$$pMI(\mathcal{T}) = \log_2 \frac{p(e_1^n)}{\prod_{i=1}^n p(e_i)}. \tag{3}$$

Since *pMI* is known to favor rare types, modifications have been proposed to weight the *pMI* estimate in order to counterbalance its low-frequency bias. Again, following Sears and Widmer (2020), we weight *pMI* by a coverage statistic that measures the proportion of compositions in the corpus that contain $\mathcal{T}$.

$$pMI_c(\mathcal{T}) = pMI(\mathcal{T}) \times \frac{|m\colon \mathcal{T} \subseteq_o m|}{|\oplus|}. \tag{4}$$

Here, $|m\colon \mathcal{T} \subseteq_o m|$ refers to the total number of compositions m in the corpus $\oplus$ that contain the ordered subset $\mathcal{T}$ from the final n-gram list at least once. Thus, the coverage statistic serves as a scaling factor for pMI_c.

The logic behind pMI_c is that it ranks patterns *within* a corpus, and so identifies *conventional* utterances that may or may not be specific to that corpus. By comparing evidence for an MCE like the compound cadence between two corpora, however, the *relative frequency ratio*, denoted by r, can be used to discover patterns that are *characteristic* of a corpus relative to a comparison corpus (Manning and Schütze. 1999).

$$r(\mathcal{T}) = \log_2 \frac{p(e_1^n \mid \oplus)}{p(e_1^n \mid \ominus)}. \tag{5}$$

Following Conklin and Bergeron (2010), $p(e_1^n \mid \oplus)$ denotes the probability of $\mathcal{T}$ in the corpus under investigation, and $p(e_1^n \mid \ominus)$ represents the probability of $\mathcal{T}$ in the comparison corpus (or *anticorpus*). Thus, a high estimate for r indicates that $\mathcal{T}$ may be characteristic of the corpus.

4. Recurrent patterns: the findings

Table 2 presents the top ten conventional 3-gram types in the ABC/TAVERN (*classical*) and RS-200/Billboard (*popular*) data sets, respectively, using IDyOM to exclude types whose terminal members do not precede a prediction error, and ranked by pMI_c, a statistical association measure that privileges patterns whose members co-occur more frequently than the background probability would suggest. For the classical corpus, the compound cadence progression received the highest ranking, reflecting its status as the most important closing pattern in Western music of the common-practice period (Meyer 2000; Sears and Widmer 2020). An incomplete variant also appeared in the top ten (Rk. 7), along with the simple cadence progression, ii^6-V^7-I (Rk. 6). The remaining 3-gram types comprise prototypical subsets of the rule of the octave or prolongational progressions of tonic harmony. Examples of the descending rule include progressions supported by the bass-line scale-degrees, Do-Te-La (Rk. 2), Sol-Fa-Mi (Rk. 3), and Mi/Me-Re-Do (Rks. 4, 8), while the ascending rule includes examples supported by the bass scale-degrees, Do-Re-Mi (Rk. 5), and La-Ti-Do (Rk. 10).

For the popular corpus, seven of the top ten types featured tonal progressions (Rks. 1, 2, 3, 4, 7, 8, 9), reinforcing claims about the influence of the major-minor diatonic system in popular music (Everett 2004; Biamonte 2010). By comparison, progressions characterized by flat-side triads, the "harmonic code [of] rock" (Biamonte 2010, 97) are notably scarce in the table; just one progression from the aeolian family appeared in the top ten (Rk. 5), while other patterns, such as the double-plagal progression, were absent. Nevertheless, patterns reflecting other diatonic systems, including the blues progression (Rk. 6), emerged near the top of a 3-gram list comprised of over 1,000 unique types after filtering. Relative to the classical corpus, the preference for root motion in the popular copora is also evident in the table. This point likely reflects the prevalence of guitar-based (rather than keyboard-based) textures in the popular corpus, where voice-leading and counterpoint are presumed to play ancillary organizational roles (Moore 2001). The scalar pattern IV^{M7}-iii^7-ii^7 is one notable example of stepwise descent in the bass that nonetheless reflects root motion. Finally, partial and/or open rotations for three of the progressions (Rks. 2, 4, 8) appeared in the top ten in spite of the selected filtering method.

Table 3 presents the top ten characteristic 3-gram types ranked by the relative frequency ratio, *r*. For the classical corpus, the compound cadence progression remained in the top ten (Rk. 4), but also featured five tokens in the popular corpus, which reduced its final ranking. Instead, other cadential and prolongational progressions emerged at the top of the table, the majority of which feature inversions that afford stepwise motion in the bass. By comparison, characteristic progressions in the popular corpus clearly reflect other diatonic systems. The closed double-plagal progression, which failed to appear in Table 2, appeared in the top ten progressions in Table 3 (Rk. 6), and also featured two open rotations (Rks. 2, 9). The closed aeolian progression also appeared in the top ten (Rk. 10), as did a partial exemplar of the tonally ambiguous axis progression (Rk. 8), and two rotations of the blues progression (Rks. 1, 7). Finally, only three tonal progressions remained in the top ten (Rks. 3, 4, 5).

Given the size of these corpora, a comparative study of the four-gram types yields patterns with much lower counts, so we did not include the tables here. The compound cadence, ii^6-V$\binom{6}{4}$-V^7-I, served as the top-ranked conventional 4-gram type in the classical corpus. Three other variants also appeared in the top ten, which raises a persistent issue in corpus studies: namely, many of the patterns are variants of the same progression. By representing each chord according to its root, quality, inversion, and extensions, the relative root encoding scheme overspecifies the underlying vocabulary in many cases, resulting in counts that are distributed between variants of the same type. In the popular corpus, for example, both the complete doo-wop progression, vi-IV-V-I, and a potential closed variant, vi-ii-V-I, appeared in the top ten.

Table 2. Top ten *conventional* 3-gram types in the ABC/TAVERN (*classical*) and RS-200/Billboard (*popular*) data sets.

	ABC/TAVERN						RS-200/Billboard					
Rk.	Type			*pMI*$_c$	*N*	Family	Type			*pMI*$_c$	*N*	Family
1	$V(^6_4)$	V^7	I	2.181	282	comp.	IV	V	I	0.348	474	simple
2	I	V^4_2/IV	IV^6	1.463	29	rule	vi	IV	V	0.138	172	doo-wop[p]
3	V	V^4_2	I^6	1.256	62	rule	ii	V	I	0.133	109	simple
4	I^6	V^4_3	I	1.211	70	rule	V	I	IV	0.117	202	simple[r]
5	I	V^4_3	I^6	0.888	46	rule	♭VI	♭VII	i	0.109	88	aeolian
6	ii^6	V^7	I	0.885	109	simple	V	IV	I	0.108	264	blues
7	V^6_5/V	$V(^6_4)$	V^7	0.864	19	comp.	ii^7	V^7	I	0.107	75	simple
8	i^6	V^4_3	i	0.826	25	rule	I	IV	V	0.105	209	simple[r]
9	I	V^4_2	I^6	0.724	49	prolong.	V^7/V	V^7	I	0.079	32	simple
10	IV^6	V^6_5	I	0.699	32	rule	IV^{M7}	iii^7	ii^7	0.074	35	scalar

Note. Family designations: *classical – simple* = simple cadence; *comp.* = compound cadence; *prolong.* = prolongational; *rule* = derived from a possible rule of the octave; *popular – dbl. plagal* = double plagal; *scalar* = diatonic root motion by step. [p] possible partial (or incomplete) progression. [r] possible open rotation.

Given the findings presented here, it seems reasonable to conclude that tonal harmonic progressions serve as stylistic conventions in both corpora. One might argue, however, that the high incidence of these progressions across the popular corpus reflects the prevalence of songs from earlier decades when progressions like the doo-wop – the so-called "fifties progression" – were particularly common (Doll 2017, 113).[9] By comparison, modal and pentatonic patterns like the double-plagal and aeolian progressions presumably appeared with much greater frequency in subsequent decades. Everett (2004) has suggested, for example, that common-practice norms eventually gave way to modal and pentatonic diatonic systems supporting flat-side harmonies, symmetric distributions, and harmonic patterns like the double-plagal, aeolian, and so on.

To examine this claim, Figure 3 tracks the incidence of the conventional (PMI_c) tonal and characteristic (r) non-tonal closed progressions in the popular corpus over time. The histogram in Figure 3(a) visualizes the proportion of songs in each year that featured the double-plagal progression from 1956 to 1991 using the unfiltered n-gram list.[10] The black line is a kernel density estimation function that smooths the histogram of proportions of songs in each year that feature the double-plagal progression in order to permit visualizations of several potential patterns (or aggregated patterns) in the same plot. In this case, the double-plagal progression emerged in 1957, peaked in 1972, and then gradually declined until 1990.

Figure 3(b) shows the smoothed functions for the histograms representing the aggregated proportions of songs in each year that featured at least one of the four conventional tonal progressions or four characteristic non-tonal progressions in the popular corpus. The tonal progressions reached their global peak in 1960, and despite a precipitous decline, remained prevalent throughout the '70s and '80s. By comparison, closed non-tonal patterns like the double-plagal, aeolian, blues, and pentatonic progressions first peaked in 1957 (due to the prevalence of the twelve-bar blues), but were generally less prevalent than their tonal counterparts for much of the fifty-year period represented in these corpora. However, from 1968 to 1973, and from 1983 to 1986, these non-tonal progressions were slightly more prevalent than the selected tonal progressions. Thus, these corpora suggest that tonal harmonic progressions remained stylistic

[9] More than half of the songs in the RS-200 data set are from the 1950s and '60s.

[10] This period featured seven or more songs in each year in the Billboard and RS-200 data sets.

Table 3. Top ten *characteristic* 3-gram types in the ABC/TAVERN (*classical*) and RS-200/Billboard (*popular*) data sets.

	ABC/TAVERN						RS-200/Billboard					
Rk.	Type			r	N	Family	Type			r	N	Family
1	ii^6	V^7	I	7.644	109	simple	V	IV	I	7.169	264	blues
2	I^6	V^4_3	I	7.005	70	rule	I	♭VII	IV	6.886	217	dbl. plagal^r
3	V	V^4_2	I^6	6.830	62	rule	I	IV	V	6.832	209	simple^r
4	$V(^6_4)$	V^7	I	6.693	282	comp.	V	I	IV	6.783	202	simple^r
5	I^6	V^6_5	I	6.604	53	prolong.	vi	IV	V	6.551	172	doo-wop^p
6	I	V^4_2	I^6	6.490	49	prolong.	♭VII	IV	I	6.201	135	dbl. plagal
7	I	V^4_3	I^6	6.399	46	rule	I	V	IV	5.825	104	blues^r
8	ii^6	V	I	6.335	44	simple	IV	I	V	5.811	103	blues^{r*}
9	IV^6	V^6_5	I	5.876	32	rule	V^7/IV	♭VII	IV	5.709	96	dbl. plagal^r
10	I	V^4_2/IV	IV^6	5.734	29	rule	♭VI	♭VII	i	5.584	88	aeolian

Note. Family designations: *classical – simple* = simple cadence; *comp.* = compound cadence; *prolong.* = prolongational; *rule* = derived from a possible rule of the octave; *popular – dbl. plagal* = double plagal. ^{p} possible partial (or incomplete) progression. ^{r} possible open rotation. *IV-I-V may also represent a partial axis progression.

conventions throughout the second half of the twentieth century, but the prevalence of non-tonal progressions also testifies to the diversity of the musical material characterizing this time period. Finally, the decline in both tonal and non-tonal progressions leading into the 1990s could result, in part, from the diminishing representation of songs from those decades in the RS-200 data set, but it might also suggest the emergence of new harmonic paradigms in the following decades.

5. Discussion

This study examined recurrent triadic patterns in the ABC, TAVERN, RS-200, and Billboard data sets. To that end, we applied methods for the discovery of multichord expressions. The analysis pipeline identified MCEs as sub-sequences from the larger sequence (n-grams), excluded tokens if their terminal members did not precede a segment boundary (IDyOM), and ranked the resulting types according to statistical association measures representing the conventional (pMI_c) and characteristic (r) features of each n-gram type. Tonal progressions appeared at the top of the final n-gram lists, suggesting they serve as stylistic conventions in the corpora selected for this study. The classical corpus also featured characteristic patterns with inverted harmonies, supporting the textural distinction between the two-voice, contrapuntal structures associated with the "classic texture" (Ratner 1980), and the guitar-based textures found in rock music. By comparison, the popular corpus featured flat-side harmonic progressions reflecting various modal and pentatonic diatonic systems, many of which subvert traditional (i.e. major-minor) notions of diatony. Finally, histograms of the most conventional and characteristic tonal and non-tonal progressions from the popular corpus revealed distinctive trends over time, with tonal progressions peaking in the mid 1960s, and non-tonal progressions peaking around 1970. The nearly equal prominence of these patterns in the second half of the twentieth century testifies to the organizational diversity of popular music.

Admittedly, this study only briefly mentioned many of the issues surrounding comparative corpus studies of this type. Perhaps the most obvious relates to the diversity of chord types reflected across the selected corpora. The cadential six-four is a textbook example. Symbols like Cc, C^6_4, Ic, ic, I^6_4, i^6_4, V^6_4, and $V(^6_4)$ all appear under the cadential six-four umbrella, even within annotations from the same encoder. Chord types without roots, such as augmented-sixth chords

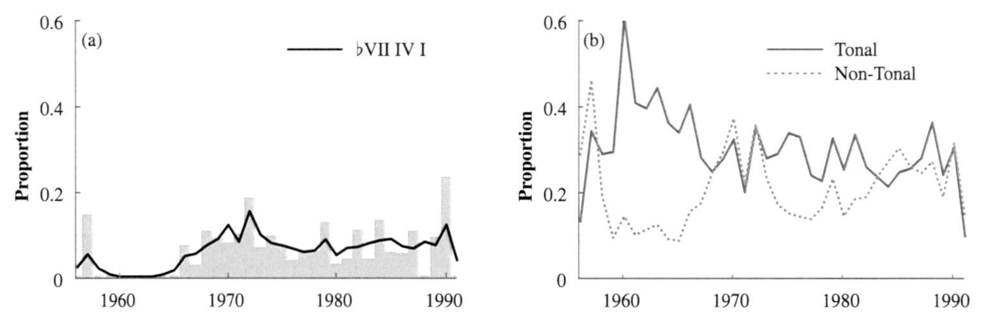

Figure 3. (a) Histogram and kernel density estimation function of the proportion of songs in each year that feature the double-plagal progression (closed) in the RS-200/Billboard data sets. (b) Kernel density estimation functions of the aggregated proportions of songs in each year that feature at least one of the top four conventional tonal and characteristic non-tonal (closed) progressions from the corresponding *n*-gram lists. Tonal – IV-V-I; ii-V-I; ii^7-V^7–I; ii^7-V^7–I^{M7}. Non-tonal – V-IV-I; ♭VI- ♭VII-i; ♭VII-IV-I; ♭III-IV-I (color).

and common-tone diminished seventh chords, also complicate efforts to employ a rigid Roman numeral encoding scheme. Perhaps worse, slash notation permits non-tertian chord tones and embellishing tones in the bass, both of which could serve as characteristic markers of popular music. Thus, comparisons with traditionally tertian corpora necessarily force the analyst to minimize potential differences in the respective vocabularies.

In addition to potential inconsistencies in the encoding scheme, the selected model pipeline operationalized certain assumptions about the relationship between statistical inference and probability theory on the one hand, and implicit learning and long-term knowledge on the other. For example, the *n*-gram lists presented here reflect a filtering method that excluded *n*-gram tokens whose terminal members preceded a peak in information content. An empirical study might instead treat this filtering method as a free parameter in order to optimize the rank of a given utterance (Sears and Widmer 2020), or evaluate the segmentation method employed here against expert annotations. For instance, it may be the case that other closing patterns will emerge at the top of an *n*-gram list simply by adjusting the segmentation threshold *k*. What is more, this study restricted its purview to 3- and 4-grams, but previous studies have developed filtering and ranking methods to identify frequent but distinctive patterns when *n* is not known in advance (Lartillot 2005; Conklin 2010). Finally, the preceding tables arbitrarily excluded MCEs beyond the top ten, many of which could still be interpreted as conventional or characteristic depending on the selected exclusion criteria.

To address these issues, we created the *Triadic Harmony Analysis Tool*, a standalone application that allows users to adjust various stages of the model pipeline and examine the resulting *n*-gram lists.[11] Shown in Figure 4, the tool features a simple graphic user interface so that users with minimal programming knowledge can examine *n*-gram lists for any combination of data sets, pattern lengths *n*, and segmentation thresholds *k*. The tool also allows users to export the resulting tables to csv files for further exploration and analysis. Thus, it is our hope that applications like this one will eventually allow researchers to examine claims for bodies of music that often far exceed the capacities of one scholarly lifetime, and in so doing, provide further evidence of the replicated patterns conventional to, and characteristic of, a given style.

[11] The analysis tool is available for download on Windows and macOS operating systems at https://osf.io/kdzm3/.

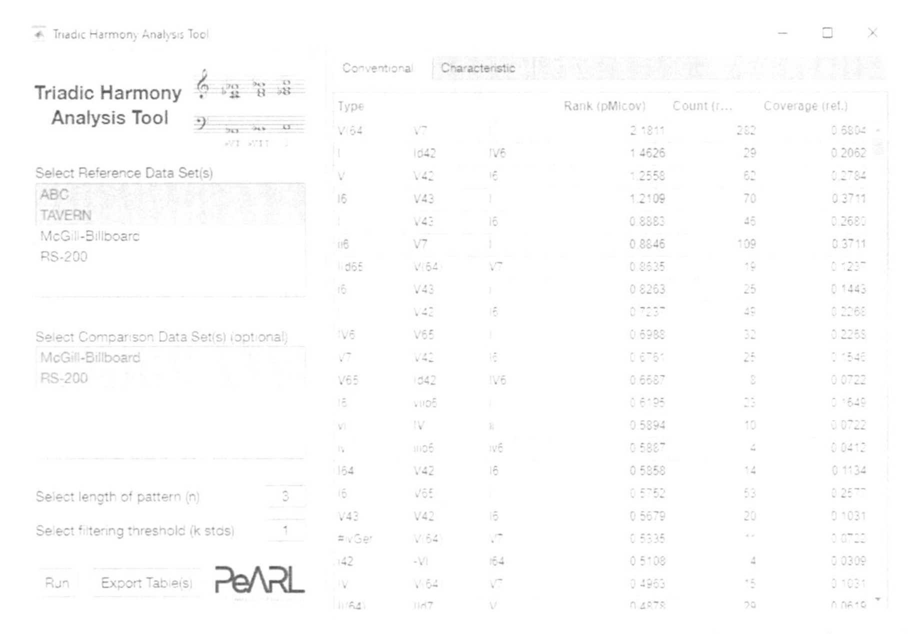

Figure 4. Screenshot of the graphic user interface for the *Triadic Harmony Analysis Tool*, a cross-platform standalone application available for download at https://osf.io/kdzm3/ (color).

Disclosure statement

No potential conflict of interest was reported by the author(s).

References

Biamonte, Nicole. 2010. "Triadic Modal and Pentatonic Patterns in Rock Music." *Music Theory Spectrum* 32 (2): 95–110.

Brent, Michael R. 1999. "An Efficient, Probabilistically Sound Algorithm for Segmentation and Word Discovery." *Machine Learning* 34 (1/3): 71–105.

Brent, Michael R. 1999. "Speech Segmentation and Word Discovery: A Computational Perspective." *Trends in Cognitive Sciences* 3 (8): 294–301.

Burgoyne, John Ashley, Jonathan Wild, and Ichiro Fujinaga. 2011. "An Expert Ground-Truth Set for Audio Chord Recognition and Music Analysis." In *Proceedings of the 12th International Society for Music Information Retrieval Conference (ISMIR)*, edited by Anssi Klapuri, and Colby Leider, Miami, FL, 423–428.

Collins, Tom, Andreas Arzt, Harald Frostel, and Gerhard Widmer. 2016. "Using Geometric Symbolic Fingerprinting to Discover Distinctive Patterns in Polyphonic Music Corpora." In *Computational Music Analysis*, edited by David Meredith, 445–474. Cham: Springer International Publishing.

Conklin, Darrell. 2010. "Distinctive Patterns in the First Movement of Brahms' String Quartet in C Minor." *Journal of Mathematics and Music* 4 (2): 85–92.

Conklin, D. 2010. "Discovery of Distinctive Patterns in Music." *Intelligent Data Analysis* 14 (5): 547–554.

Conklin, Darrell, and Mathieu Bergeron. 2008. "Feature Set Patterns in Music." *Computer Music Journal* 32 (1): 60–70.

Conklin, Darrell, and Mathieu Bergeron. 2010. "Discovery of Contrapuntal Patterns." In *Proceedings of the 11th International Society for Music Information Retrieval (ISMIR)*, Utrecht, The Netherlands, 201–206.

Declercq, Trevor, and David Temperley. 2011. "A Corpus Analysis of Rock Harmony." *Popular Music* 30 (1): 47–70.

Devaney, Johanna, Claire Arthur, Nathaniel Condit-Schultz, and Kirsten Nisula. 2015. "Theme and Variation Encodings with Roman Numerals (TAVERN): A New Data Set for Symbolic Music Analysis." In

Proceedings of the 16th International Society for Music Information Retrieval (ISMIR), Málaga, Spain, 728–734.

Doll, Christopher. 2017. *Hearing Harmony: Toward a Tonal Theory for the Rock Era*. Ann Arbor, MI: University of Michigan Press.

Everett, Walter. 2004. "Making Sense of Rock's Tonal Systems." *Music Theory Online* 10 (4). http://mto.societymusictheory.org/issues/mto.04.10.4/mto.04.10.4.w_everett.html.

Evert, Stefan. 2008. "Corpora and Collocations." In *Corpus Linguistics: An International Handbook*, edited by Anke Lüdeling and Merja Kytö, article 58. Berlin: Mouton de Gruyter.

Gjerdingen, Robert O. 2007. *Music in the Galant Style: Being an Essay on Various Schemata Characteristic of Eighteenth-Century Music*. New York: Oxford University Press.

Harte, Christopher, Mark Sandler, Samer Abdallah, and Emilia Gómez. 2005. "Symbolic Representation of Musical Chords: A Proposed Syntax for Text Annotations." In *Proceedings of the 6th International Society for Music Information Retrieval Conference (ISMIR)*, London, UK, 66–71.

Huron, David. 2001. "What is a Musical Feature? Forte's Analysis of Brahms's Opus 51, No. 1, Revisited." *Music Theory Online* 7 (4). https://www.mtosmt.org/issues/mto.01.7.4/mto.01.7.4.huron.html.

Hyer, Brian. 2002. "Tonality." In *Cambridge History of Western Music Theory*, edited by Thomas Christensen, 726–752. Cambridge: Cambridge University Press.

Jusczyk, Peter W. 1997. *The Discovery of Spoken Language*. Cambridge, MA: MIT Press.

Kurby, Chistopher A., and Jeffrey M. Zacks. 2008. "Segmentation in the Perception and Memory of Events." *Trends in Cognitive Sciences* 12 (2): 72–79.

Lartillot, Olivier. 2005. "Efficient Extraction of Closed Motivic Patterns in Multi-Dimensional Symbolic Representations of Music." In *Proceedings of the 2005 IEEE/ACM International Conference on Web Intelligence (WI'05)*, Compiegne, France, 229–235.

Manning, Christopher D., and Hinrich Schütze. 1999. *Foundations of Statistical Natural Language Processing*. Cambridge, MA: MIT Press.

Meyer, Leonard B. 1957. "Meaning in Music and Information Theory." *The Journal of Aesthetics and Art Criticism* 15 (4): 412–424.

Meyer, Leonard B. 1989. *Style and Music: Theory, History, and Ideology*. Philadelphia, PA: University of Philadelphia Press.

Meyer, Leonard B. 2000. "Nature, Nurture, and Convention: The Cadential Six-Four Progression." In *The Spheres of Music*, 226–263. Chicago: The University of Chicago Press.

Moore, Allan F. 2001. *Rock: The Primary Text – Developing a Musicology of Rock*. 2nd ed. Aldershot: Ashgate Publishing.

Moss, Fabian C., Markus Neuwirth, Daniel Harasim, and Martin Rohrmeier. 2019. "Statistical Characteristics of Tonal Harmony: A Corpus Study of Beethoven's String Quartets." *PLoS ONE* 14 (6): e0217242.

Narmour, Eugene. 1990. *The Analysis and Cognition of Basic Melodic Structures: The Implication-Realization Model*. Chicago, IL: University of Chicago Press.

Neuwirth, Markus, Daniel Harasim, Fabian C. Moss, and Martin Rohrmeier. 2018. "The Annotated Beethoven Corpus (ABC): A Dataset of Harmonic Analyses of All Beethoven String Quartets." *Frontiers in Digital Humanities* 5 (16).

Pearce, Marcus T. 2005. "The Construction and Evaluation of Statistical Models of Melodic Structure in Music Perception and Composition." Unpublished doctoral dissertation, City University, London, London, UK.

Pearce, Marcus T., Daniel Müllensiefen, and Geraint A. Wiggins. 2010. "The Role of Expectation and Probabilistic Learning in Auditory Boundary Perception: A Model Comparison." *Perception* 39 (10): 1367–1391.

Pecina, Pavel. 2005. "An Extensive Empirical Study of Collocation Extraction Methods." In *Proceedings of the ACL Student Research Workshop*, Ann Arbor, MI, 13–18.

Ratner, Leonard G. 1980. *Classic Music: Expression, Form, and Style*. New York: Schirmer Books.

Richards, Mark. 2017. "Tonal Ambiguity in Popular Music's Axis Progressions." *Music Theory Online* 23 (3). https://mtosmt.org/issues/mto.17.23.3/mto.17.23.3.richards.html

Saffran, Jenny R. 2003. "Statistical Language Learning: Mechanisms and Constraints." *Current Directions in Psychological Science* 12 (4): 110–114.

Sanguinetti, Giorgio. 2012. *The Art of Partimento: History, Theory, and Practice*. Oxford, UK: Oxford University Press.

Sears, David R. W., Marcus T. Pearce, William E. Caplin, and Stephen McAdams. 2018. "Simulating Melodic and Harmonic Expectations for Tonal Cadences Using Probabilistic Models." *Journal of New Music Research* 47 (1): 29–52.

Sears, David R. W., and Gerhard Widmer. 2020. "Beneath (or Beyond) the Surface: Discovering Voice-Leading Patterns with Skip-Grams." *Journal of Mathematics and Music*. https://doi.org/10.1080/17459737.2020.1785568.

Shanahan, Daniel, and Yuri Broze. 2012. "A Diachronic Analysis of Harmonic Schemata in Jazz." In *Proceedings of the 12th International Conference on Music Perception and Cognition and the 8th Triennial Conference of the European Society for the Cognitive Sciences of Music,* edited by Emilios Cambouropoulos, Costas Tsougras, Panayotis Mavromatis, and Konstantinos Pastiadis, Thessaloniki, Greece, 909–917. Aristotle University of Thessaloniki.

Tymoczko, Dmitri. 2011. *A Geometry of Music*. Oxford: Oxford University Press.

White, Christopher Wm. 2015. "A Corpus-Sensitive Algorithm for Automated Tonal Analysis." In *Mathematics and Computation in Music*, edited by Tom Collins, David Meredith, and Anja Volk, 115–121. Cham, Switzerland: Springer International Publishing.

Modelling pattern interestingness in comparative music corpus analysis

Kerstin Neubarth and Darrell Conklin

In computational pattern discovery, pattern evaluation measures select or rank patterns according to their potential interestingness in a given analysis task. Many measures have been proposed to accommodate different pattern types and properties. This paper presents a method and case study employing measures for frequent, characteristic, associative, contrasting, dependent, and significant patterns to model pattern interestingness in a reference analysis, Frances Densmore's study of Teton Sioux songs. Results suggest that interesting changes from older to more recent Sioux songs according to Densmore's analysis are best captured by contrast, dependency, and significance measures.

1. Introduction

Pattern discovery provides powerful and versatile techniques for symbolic music analysis. Patterns in music include *intra-opus patterns*, repeated within a single piece of music, and *inter-opus patterns*, occurring across multiple pieces in a music corpus (Conklin 2010a). Inter-opus pattern discovery can be applied to unstructured corpora, extracting patterns that describe general features of the represented repertoire (e.g. Conklin and Anagnostopoulou 2001), or to partitioned corpora, extracting patterns that distinguish classes of music pieces such as different song types, geographic regions, or composers (e.g. Conklin and Anagnostopoulou 2011; Collins et al. 2016). Traditionally, pattern mining in music – both intra- and inter-opus analysis – has been dominated by work on discovering *sequential patterns*. For inter-opus pattern mining, *global-feature patterns* offer an alternative pattern representation (e.g. Taminau et al. 2009; Shanahan, Neubarth, and Conklin 2016). This paper studies inter-opus, global-feature patterns in class-labelled music corpora.

Searching for patterns in music may proceed as *deductive* analysis, which retrieves instances of specified patterns, or as *inductive* analysis, which finds unspecified patterns satisfying certain criteria of pattern interestingness (Conklin 2010b). Beyond music data mining, many measures for quantifying pattern interestingness have been proposed, originating from different contexts including statistics and information theory (Geng and Hamilton 2006); relating to different pattern types, such as frequent or contrasting patterns (Dong and Li 1999; Bay and Pazzani

2001; Han et al. 2007); and satisfying different properties regarding, e.g. their scaling behaviour (Piatetsky-Shapiro 1991; Tan, Kumar, and Srivastava 2002; Lenca et al. 2007). Interestingness measures can be employed to distinguish interesting from uninteresting patterns, usually requiring the definition of a measure threshold, or to rank patterns. Measures are used during pattern discovery to prune the search space, or during post-processing to filter or rank the output of discovered patterns (Geng and Hamilton 2006).

The work presented in this paper explores computational measures for modelling the interestingness of patterns suggested by extant music analyses. Hence it lies at the intersection of deductive and inductive analysis: it shares with deductive analysis the study of given patterns, and for studying these patterns makes use of pattern evaluation criteria usually employed in inductive analysis. More specifically, we report a case study on patterns in Native American music: a meta-analysis of Frances Densmore's analysis of Teton Sioux music, which investigates changes from older to more modern Sioux songs (Densmore 1918). The case study illustrates core interests in data mining (e.g. Dong and Li 1999) and computational music corpus analysis (e.g. Jackson 1970; Broze and Shanahan 2013): discovering changes in chronologically structured data.

2. Reference analysis: interesting changes in Teton Sioux music

Frances Densmore (1867–1957) was one of the most prolific collectors of North American native music. The case study in this paper focuses on Densmore's analysis of Teton Sioux songs, collected on the Standing Rock and Sisseton Reservations in North and South Dakota between 1911 and 1914 (Densmore 1918). Like most of her publications with the Bureau of American Ethnology, the study of Teton Sioux music includes quantitative analyses of the documented songs based on global music content features, which capture the "melodic trend and general musical character" of the songs (Densmore 1910, 3).

2.1. *Densmore's collection of Teton Sioux music*

Densmore's publication on Teton Sioux music presents transcriptions and analyses of 240 songs. The corpus is organised according to "the age of the songs, this series being divided for analysis into two groups, one comprising songs believed to be more than 50 years old and the other comprising songs of more recent origin" (Densmore 1918, v). The class of older songs contains 147 songs associated with obsolete ceremonies or recorded by old men who had learned or received the songs in their youth. The class of comparatively modern songs comprises 93 songs recorded by young men, linked to modern tribal societies, or referencing a recent custom.

To describe musical properties of songs Densmore applied global features, i.e. song-level attribute–value pairs, which capture melodic and rhythmic-metric aspects of songs (Table 1). For the computational analysis, we collated the feature encoding for the 240 songs from Densmore's publication. To allow comparison with Densmore's analysis results, we aggregated attribute values when suggested by Densmore's textual description. For example, with respect to the attribute compass Densmore commented on songs "having a range of 12 or more tones" (Densmore 1918, 24), aggregating values from 12 to 17 tones. For comparing the tempo of old and more modern Sioux songs, the values of attributes tempoVoice and tempoDrum (measured in metronome values) were aggregated into two bins, with a split point at the median, reducing the 30 resp. 27 fragmented and infrequently observed attribute values considered in Densmore's original tabulated analysis to two categorical values slow, covering metronome markings 48–96, and

Table 1. Music content descriptors in Densmore's analysis of Teton Sioux songs (Densmore 1918).

Attribute	Description
tonality	Tonality [according to major/minor third above keynote]
firstReKey	First note of song – its relation to keynote
lastReKey	Last note of song – its relation to keynote
lastReCompass	Last note of song – its relation to compass of song
compass	Number of tones comprising compass of song
material	Tone material
accidentals	Accidentals [chromatic alterations of tones]
structure	Melodic structure [relation between contiguous accented tones]
firstProgression	First progression – downward and upward
firstMetricPos	Part of measure on which song begins
firstMeasure	Rhythm (metre) of first measure
metreChange	Change of time (measure-lengths)
rhythmDrum	Rhythm of drum
rhythmicUnit	Rhythmic unit of song
tempoVoice	Metric unit of voice (tempo)
tempoDrum	Metric unit of drum (tempo)
tempoVoiceDrum	Comparison of metric unit of voice and drum (tempo)

firstProgression : up
firstMetricPos : accented
firstMeasure : 2-4
metreChange : yes
rhythmDrum : quarter_unaccented
rhythmicUnit : one
tempoVoice : slow
tempoDrum : slow
tempoVoiceDrum : same

Figure 1. Densmore's transcription of the modern Teton Sioux song "Song of the Buffalo Hunt (c)" (Cat. No. 545, Densmore 1918, 442) with encoding by selected global features. Angular brackets mark the rhythmic unit.

fast, covering metronome markings 100–192. Figure 1 shows a short example song encoded by selected global features.

2.2. *Reference patterns in Densmore's analysis of Teton Sioux music*

Densmore's comparative analysis of old and modern Sioux songs studied one attribute at a time, i.e. the analysis reveals single-feature patterns. To identify reference patterns in Densmore's analysis, we first extracted all pairs ⟨*feature*, *class*⟩ which are mentioned in the textual descriptions accompanying Densmore's quantitative analyses. From these, redundant patterns – due to symmetries between the two classes or in cases of only two attribute values – were removed. For example, the "larger proportion [of songs] having a range of 12 or more tones" (Densmore 1918, 24) among the old songs implies a smaller proportion of wide-range songs in the modern group (Densmore 1918, 25), and a "decrease in the percentage of songs having a change of measure-lengths" from the old to the modern songs implies "an increase [. . .] in songs without change in time" (Densmore 1918, 25). Of such symmetric pairs, we retained the over-represented pattern, in the above examples ⟨compass : twelve_or_more,old⟩ and ⟨metreChange : no,modern⟩. Where the text mentions only the under-represented pattern, for the current study this was replaced by the corresponding over-represented pattern, e.g. "[t]he modern songs show a smaller proportion of songs in which the final tone is the lowest in the song" (Densmore 1918, 24) was recorded as the complementary group of old songs containing a larger proportion of songs ending on the lowest tone, i.e. by the pattern ⟨lastReCompass : lowest,old⟩.

In a second step, we assigned the reference patterns a level of interestingness based on Densmore's description. Analysing changes from older to newer Teton Sioux songs, Densmore appears to distinguish different degrees of change: for example, "the newer group shows an increase in the proportion of songs which begin in 2-4 time" (Densmore 1918, 24) but "a *large* increase in the proportion having two or more rhythmic units" (Densmore 1918, 25, our emphasis). Additionally, Densmore's comments on the tabulated analyses are followed by a concluding paragraph – "[s]ummarizing briefly the results of a comparison of the old and the more modern Sioux songs" (Densmore 1918, 25) – which recapitulates a selection of the previously presented observations. Assuming the patterns highlighted in the summary to be of particular interest, we derived an ordinal scale of four *reference levels* of interestingness:

A: patterns covered in the summary and described, in the preceding text, by a qualifier marking a pronounced (e.g. "large" or "decided") change;

B: patterns which are included in the summary but have not been qualified as a pronounced change;

C: patterns mentioned in the text but not in the summary (none of which is explicitly qualified as a pronounced change);

D: patterns which show "no material differences" (Densmore 1918, 25) or for which the proportion "is the same in the two groups" (Densmore 1918, 23).

The analysis results in 33 reference patterns (see Figure 3 in Section 3.2): six patterns at level **A**, five at level **B**, four at level **C**, and 18 patterns (nine patterns potentially associated with either class) at level **D**.

3. Computational analysis of the reference patterns

To model pattern interestingness in Densmore's analysis, the reference patterns are ranked by computational pattern interestingness measures, considering measures for different types of patterns: frequent and characteristic, contrasting, associative, dependent, and significant patterns.

3.1. *Pattern interestingness measures*

The pattern interestingness measures considered in this paper evaluate the distribution of pattern occurrences in a partitioned corpus. They map observed frequencies onto a numeric interestingness value, computed from a 2×2 contingency table (Figure 2). Here, the variables C and X refer to predicates on songs, the class and pattern predicate, respectively. A song satisfies a class C if it is annotated with the corresponding class label; a song satisfies a global-feature pattern X if the song has the attribute value indicated by the global feature; a song satisfies the conjunction of the class and pattern predicates, denoted $X \wedge C$, if it satisfies both X and C. We refer to the songs satisfying a specific class C as the *target class* of analysis and the songs not satisfying C as the *background*. To denote observed frequencies, the following notation is used: $n(C)$ is the number of songs in the target class, while $n(\neg C)$ is the number of songs in the background; $n(X)$ is the number of songs in the corpus satisfying the global-feature pattern X, while $n(\neg X)$ is the number of songs in the corpus which do not satisfy pattern X. Further, $n(X \wedge C)$ is the number of songs in the target class which satisfy pattern X, $n(\neg X \wedge C)$ the number of songs in the target class which do not satisfy pattern X, $n(X \wedge \neg C)$ the number of songs in the background which satisfy pattern X, and $n(\neg X \wedge \neg C)$ the number of songs in the background which do not satisfy pattern X. The variable N denotes the total number of songs

	C	$\neg C$	
X	$n(X \wedge C)$	$n(X \wedge \neg C)$	$n(X)$
$\neg X$	$n(\neg X \wedge C)$	$n(\neg X \wedge \neg C)$	$n(\neg X)$
	$n(C)$	$n(\neg C)$	N

Figure 2. Contingency table for a pattern $\langle X, C \rangle$.

Table 2. Interestingness measures for patterns $\langle X, C \rangle$.

Pattern type	Measure	Definition
Frequent	Coverage	$P(X)$
	Support	$P(X \wedge C)$
Characteristic	Sensitivity	$P(X \mid C)$
	IC^{++}	$\begin{cases} P(C)\left[1 - \frac{P(\neg X \mid C)}{P(\neg X \mid \neg C)}\right] & \text{if } 0 \leq \frac{P(\neg X \mid C)}{P(\neg X \mid \neg C)} < 1 \\ 0 & \text{otherwise} \end{cases}$
Contrasting	Support difference	$P(X \mid C) - P(X \mid \neg C)$
	Growth rate	$P(X \mid C) \ / \ P(X \mid \neg C)$
Associative	Confidence	$P(C \mid X)$
Dependent	PS	$P(X \wedge C) - P(X)P(C)$
	Interest	$P(X \wedge C) \ / \ P(X)P(C)$
	Conviction	$P(X)P(\neg C) \ / \ P(X \wedge \neg C)$
Significant	p-value (Fisher)	$P_F(X, C)$

in the corpus. The relative frequency of the pattern in the corpus, or its empirical probability, is $P(X) = n(X)/N$, and the relative frequency of the pattern in a class, or its conditional probability given the class, is $P(X|C) = n(X \wedge C)/n(C)$. Building on these definitions, Table 2 lists probability-based measures for frequent and characteristic, contrasting, associative, dependent, and significant patterns.

3.1.1. *Frequent and characteristic patterns*

Frequent pattern mining is a core task in both wider data mining and music data mining. Patterns are considered frequent if they occur in a data set with frequency above a user-specified threshold (Han et al. 2007). In frequent pattern mining of class-labelled corpora, *coverage* measures the relative frequency of a pattern in a corpus, while *support* measures the relative co-occurrence of a pattern and a specific class (Geng and Hamilton 2006). *Sensitivity* computes the proportion of instances in a class that satisfy a pattern (Lavrač, Flach, and Zupan 1999). A pattern which is shared by all or most instances in a class is considered characteristic of the class (Han et al. 1996). Alternative measures prefer characteristic patterns which are also distinctive of the class. As an example, this study includes the IC^{++} measure (Kamber and Shinghal 1996): the more instances in the target class, relative to the background, do not satisfy the pattern, the less characteristic of the class is the pattern.

3.1.2. *Contrasting patterns*

Contrast pattern mining identifies differences between classes in categorically partitioned data or trends in chronologically partitioned corpora (Dong and Li 1999; Bay and Pazzani 2001). Measures for contrasting patterns generally compare the observed relative frequencies of a pattern in a target class and in the background. The measure of *support difference*, used in contrast set mining (Bay and Pazzani 2001), calculates their difference, while *growth rate*, used in emerging pattern mining (Dong and Li 1999), computes their ratio. Although Densmore does not systematically quantify differences in her tabulated analyses, for both support difference and growth rate corresponding examples can occasionally be found in her textual description. As an example of the first, her comparison of old and modern Sioux songs shows that "the proportion beginning on the octave is 10 per cent greater in the modern songs" (Densmore 1918, 23–24). On the other hand, her analysis of Sioux, Chippewa, and Ute songs gives an example of growth rate: "The percentage of songs of a mixed form is more than twice as great in the Ute as in the Chippewa and Sioux" (Densmore 1922, 53).

3.1.3. *Associative patterns*

Associations describe frequently co-occurring patterns (Han et al. 1996); class associations relate frequent patterns to classes (Liu, Hsu, and Ma 1998). Classic methods for mining class association patterns combine the measures of support to ensure sufficiently frequent patterns and *confidence* to assess the strength of the association between a pattern and a class: confidence corresponds to the conditional probability of the class given the pattern (Liu, Hsu, and Ma 1998). The confidence measure does not take into account the prior probability of the class, thus a class association pattern may be confident when pattern and class are not correlated or even negatively correlated (e.g. Brin, Motwani, and Silverstein 1997).

3.1.4. *Dependent patterns*

To address problems of the confidence measure, alternative measures have been used, including the *PS* measure (Piatetsky-Shapiro 1991), *interest* (Brin, Motwani, and Silverstein 1997), and *conviction* (Brin et al. 1997): the PS measure, by taking the difference, and interest, by calculating the ratio, compare the joint and individual probabilities of the pattern and the target class, indicating to which extent pattern and class are statistically dependent. Conviction considers the pattern's occurrence in the background to quantify the dependence between pattern and target class. Hence, while the contrast measures compare the pattern's observed count in the target class against its observed count outside the class, relative to the size of the class and the background, the dependency measures compare the pattern's observed count in the class against its count expected under the assumption of pattern and class being independent. Rewriting the measures with absolute rather than relative frequencies – by multiplying the two summands or factors by N to give $n(X \wedge C)$ and $N \times P(X) \times P(C)$ for PS and interest resp. $n(X \wedge \neg C)$ and $N \times P(X) \times P(\neg C)$ for conviction – makes explicit the comparison.

3.1.5. *Significant patterns*

Statistical significance tests estimate the likelihood of encountering observed pattern frequencies due to chance alone (Webb 2007). Tests such as Fisher's test have been applied in class association mining both in combination with other measures, such as growth rate, confidence or PS, and on their own (e.g. Conklin 2013; Shanahan, Neubarth, and Conklin 2016; Li and Zaiane 2017; Neubarth, Shanahan, and Conklin 2018). The *p-value* computed by Fisher's exact test (right tail) gives the probability of observing $n(X \wedge C)$ or more occurrences of the pattern in the target class given the marginal counts $n(X)$, $n(C)$ and N. The lower the p-value, the more interesting is the pattern.

3.2. *Comparison of reference and computational pattern interestingness*

The interestingness measures listed in Table 2 were applied for computational evaluation of the reference patterns, adjusting marginal counts in the contingency table for missing values (Neubarth, Shanahan, and Conklin 2018). For each interestingness measure, the evaluated patterns were ranked from highest to lowest measure value; in the case of Fisher's test, lower p-values indicate a higher degree of interestingness. In a second step, the computationally ranked patterns were mapped onto ordinal levels of interestingness, following the procedure of Ohsaki et al. (2004): based on the categorisation of the reference patterns (6 patterns at level **A**, 5 patterns at level **B**, 4 patterns at level **C**, and 18 patterns at level **D**; see Figure 3, column "Densmore"), for each interestingness measure the six most highly ranked patterns were assigned the interestingness level **A**, the next five patterns were assigned **B**, the following four patterns were assigned **C**, and the remaining patterns were assigned **D**. This mapping then provides a basis for comparing interestingness ratings of human and computational analysis qualitatively, by visualisation, or quantitatively, by determining the number of reference patterns matched in their interestingness levels by the computational evaluation (Ohsaki et al. 2004).

Figure 3 presents the interestingness ratings, for each reference pattern, based on Densmore's analysis (column "Densmore") and assigned by the computational pattern interestingness measures (columns "p-value" to "coverage"). Patterns are grouped into the reference levels suggested from Densmore's analysis, from **A** (top) to **D** (bottom). Rows within levels are ordered according to agreement across measures, while columns are sorted by agreement across patterns. Comparing computationally evaluated pattern interestingness against the reference levels extracted from Densmore's analysis, the measures roughly fall into two groups. The measures for frequent patterns – coverage, support, and sensitivity – are largely unsuccessful in distinguishing uninteresting from interesting patterns, according to the reference analysis, matching only ten out of 18 patterns at reference level **D** and at most two out of five patterns at reference level **A**. A larger number of corresponding ratings is achieved by the IC^{++} measure, which is biased towards characteristic patterns that also distinguish the target class from other classes (Kamber and Shinghal 1996). On the other hand, measures for contrasting and for dependent and significant patterns show overall high agreement with reference ratings, with p-value, support difference and PS matching reference levels slightly better (all except one reference pattern matched at level **A** and all reference patterns matched at level **D**) than growth rate, conviction and interest (three – in the case of interest two – out of five patterns matched at level **A** and 17 out of 18 patterns matched at level **D**). At levels **B** and **C**, for several patterns the ratings are reversed with respect to the reference ratings, suggesting that the inclusion of a pattern in Densmore's summary may be less indicative of quantitative interestingness ratings than her differentiation between strong and neutral changes (e.g. "large increase" vs. "increase"), and that Densmore's selection of patterns included in the summary may be partly based on other, musical or contextual, rather than statistical considerations.

3.3. *Analysis of interestingness ratings by computational measures*

This section further analyses the differences in interestingness ratings by different computational measures, discussing selected patterns (Table 3). To generalise observations beyond example patterns, we refer to established measure properties which define a measure's behaviour for varying contingency tables. In particular, we build on two well-known properties (Piatetsky-Shapiro 1991; Tan, Kumar, and Srivastava 2002): the first covers two scenarios affecting the relative pattern frequencies in the target class and background, the second supports comparison of interestingness ratings for frequent and infrequent patterns, and for different class distributions (Table 4).

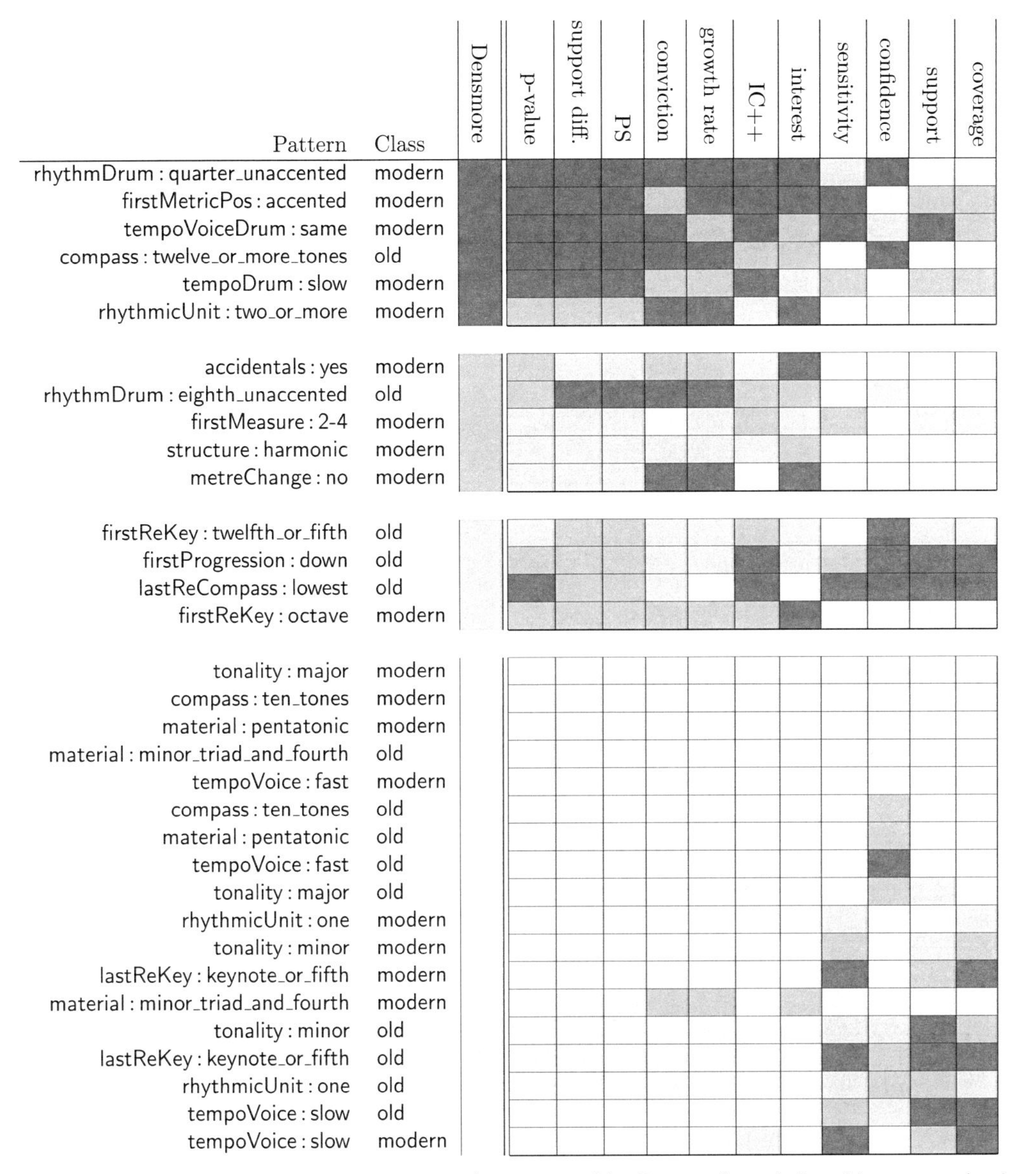

Figure 3. Comparison of pattern interestingness ratings suggested by Densmore's analysis and by computational interestingness measures. Colour legend (interestingness levels): level **A**; level **B**; level **C**; level **D**.

3.3.1. *Frequency measures*

Of the measures for frequent and characteristic patterns, support and sensitivity do not consider the background, while coverage quantifies pattern frequency across all classes in a corpus. Hence, sensitivity (coverage) assigns high ranks to patterns that are frequent in the target class (corpus) even if they are similarly or more frequent in the background. For example, the feature lastReKey : keynote_or_fifth occurs in 84% of the old and 85% of the modern songs (see Table 3): both associations are ranked among the top-6 patterns (level **A**) by sensitivity and coverage, while they are ranked in agreement with the reference analysis (level **D**) by measures that compare pattern occurrence in the target class and in the background, such as support difference

Table 3. Distribution of selected patterns in old and modern Teton Sioux songs (ordered according to their reference in the text).

	old		modern	
	$n(X \wedge C)$	$P(X \mid C)$	$n(X \wedge C)$	$P(X \mid C)$
lastReKey : keynote_or_fifth	124	0.84	79	0.85
compass : twelve_or_more_tones	47	0.32	20	0.22
rhythmicUnits : two_or_more	14	0.09	16	0.17
firstReKey : twelfth_or_fifth	63	0.43	33	0.35
firstMetricPos : accented	75	0.51	69	0.74
tempoDrum : slow*	31	0.52	43	0.69
metreChange : no	9	0.06	9	0.10
firstReKey : octave	30	0.20	27	0.29
lastReCompass : lowest	133	0.90	76	0.82

The pattern marked by an asterisk was evaluated taking into account missing values (87 of the old and 31 of the modern songs were recorded without drum, cf. Densmore 1918, 21, footnote 1).

Table 4. Properties of interestingness measures (for details see text).

		Measure properties			
Pattern type	Measure	M1.1	M1.2	M2.1	M2.2
Frequent	Coverage	Increases	Constant	Increases	Constant
	Support	Constant	Constant	Increases	Increases
Characteristic	Sensitivity	Constant	Decreases	Increases	Constant
	IC^{++}	Decreases	Decreases	Increases	Increases
Contrasting	Support difference	Decreases	Decreases	Increases	Increases*
	Growth rate	Decreases	Decreases	Constant	Increases*
Associative	Confidence	Decreases	Constant	Constant	Increases
Dependent	PS	Decreases	Decreases	Increases	Increases*
	Interest	Decreases	Decreases	Constant	Constant
	Conviction	Decreases	Decreases	Constant	Increases*
Significant	p-value	Decreases†	Decreases†	Increases†	Increases†

Notes: M1.1, change with increasing $P(X)$ when $P(X \wedge C)$ and $P(C)$ remain the same; M1.2, change with increasing $P(C)$ when $P(X \wedge C)$ and $P(X)$ remain the same; M2.1, change with scaling the first row of the contingency table by a positive factor; M2.2, change with scaling the first column of the contingency table by a positive factor.

*Direction of change for over-represented patterns, i.e. for $P(X \mid C) > P(X \mid \neg C)$; opposite direction of change for under-represented patterns.

†For p-value, with lower values indicating higher interestingness, entries refer to change in interestingness.

and growth rate, as well as IC^{++}. On the other hand, coverage, sensitivity, and support penalise infrequent features, such as compass : twelve_or_more_tones or rhythmicUnits : two_or_more (assigning level **D**), even if they are distinctive for one of the classes (listed at reference level **A** and assigned levels **A** or **B** by contrast, dependency, and significance measures).

Formally, these observations are captured by measure property M1.1 in Table 4. The property describes a measure's behaviour when, given the same pattern frequency in the target class, a higher number of pattern occurrences is observed in the background: with increasing pattern frequency in the corpus, $n(X)$, but constant pattern frequency in the target class, $n(X \wedge C)$, more pattern occurrences are found in the background, both in absolute frequency $n(X \wedge \neg C)$ and – with constant $P(C)$ and therefore $P(\neg C)$ – in relative frequency $P(X|\neg C)$. Hence, for patterns over-represented in the target class the difference between a pattern's frequency in the class and in the background decreases. Contrast measures, as well as dependency and significance measures, accordingly decrease their value. Support and sensitivity, on the other hand, remain constant while coverage increases (Table 4, column M1.1).

3.3.2. *Association measures*

Confidence quantifies the proportion of pattern occurrences observed in the target class, without taking into account the prior probability of the class (Brin, Motwani, and Silverstein 1997). For illustration, pattern ⟨firstReKey : twelfth_or_fifth,old⟩, included at reference level **C**, is ranked higher by confidence (level **A**) than by the other measures. Of the 96 songs which begin on the twelfth or fifth above the keynote, 63 songs are old songs (see Table 3), giving a confidence of 66%, which is only slightly above the proportion of old songs in the corpus (61%). On the other hand, pattern ⟨firstMetricPos : accented,modern⟩, listed at reference level **A**, is ranked lower by confidence (level **D**) than any other measure, despite its confidence being higher than expected given the prior probability of the class modern (39%): of the 144 songs starting with an accented tone, 69 songs are modern songs (see Table 3), giving a confidence of 48%. In terms of change in relative frequency from background to target class, the beginning on the twelfth or fifth above the keynote shows a difference of only 8% (see Table 3: 0.35 to 0.43), leading to the lower ranking by contrast as well as dependency and significance measures. In comparison, the beginning with an accented note increases by 23% from older to more modern songs (see Table 3: 0.51 to 0.74), leading to the higher ranking by contrast as well as dependency and significance measures. More generally, most of the patterns ranked high by confidence relate to class old.

Confidence differs from measures for contrasting, dependent, and significant patterns with respect to measure property M1.2 in Table 4. Here the absolute numbers of pattern occurrences in the target class and in the background remain the same, but their relative frequencies change: with increasing $n(C)$, and therefore decreasing $n(\neg C)$, the pattern's relative frequency in the target class $P(X \mid C)$ decreases, while its relative frequency in the background $P(X \mid \neg C)$ increases. Thus again the degree to which a pattern is over-represented in the target class relative to the background decreases. While contrast, dependency, and significance measures decrease, confidence remains constant (Table 4, column M1.2).

3.3.3. *Contrast measures*

Support difference and growth rate directly compare pattern occurrence in the target class and in the background, thus they capture changes in relative pattern frequency between older and more modern Sioux songs. The two measures differ in their ranking of frequent and infrequent patterns: for infrequent patterns a clear increase in relative frequency measured as a ratio (growth rate) is easier to achieve than when measured as a difference (support difference). In fact, contrast pattern discovery employing growth rate is specifically designed for also detecting changes in data when patterns are rare (Dong and Li 1999). For example, pattern ⟨tempoDrum : slow,modern⟩ is ranked higher by support difference (level **A**) than by growth rate (level **B**), while pattern ⟨metreChange : no,modern⟩ is ranked higher by growth rate (level **A**) than by support difference (level **C**). The feature tempoDrum : slow describes 69% of the modern and 52% of the old songs which are recorded with drum (see Table 3). On the other hand, the feature metreChange : no describes only 18 songs (7.5%) in the corpus, with nine songs in each of the two classes (6% and 10% respectively in the old and in the modern songs, see Table 3).

This difference between support difference and growth rate can be formally described with reference to measure property M2.1 in Table 4. Scaling the first row of the contingency table by a positive factor increases a pattern's frequency in the corpus but preserves the ratio between $P(X \mid C)$ and $P(X \mid \neg C)$. Growth rate therefore remains constant; support difference, on the other hand, increases for more frequent patterns (Table 4, column M2.1). In other words, for the same growth rate value a larger difference between $P(X|C)$ and $P(X \mid \neg C)$ is required at higher pattern frequencies.

3.3.4. *Dependency and significance measures*

Among the measures for dependent and significant patterns, the interest and conviction measures are more sensitive to deviations in infrequent patterns than the PS measure and p-value. Rewriting PS as weighted relative accuracy, $P(X) \times [P(C \mid X) - P(C)]$, more obviously exposes the measure's bias towards frequent patterns: in subgroup discovery, weighted relative accuracy is applied to discover distinctive patterns which are as frequent as possible (Kavšek and Lavrač 2006). Accordingly, the pattern ⟨lastReCompass : lowest,old⟩ – involving feature lastReCompass : lowest observed for 87% of all songs (see Table 3: 133 old and 76 modern songs, that is 209 of 240 songs) – is ranked higher by the PS measure (level **B**) than by conviction (level **C**) and interest (level **D**). On the other hand, the infrequent pattern ⟨metreChange : no,modern⟩ is ranked higher by conviction and interest (level **A**) than by PS (level **C**). In turn, conviction and interest differ in their interestingness ratings for different class sizes: in mining imbalanced data sets, interest has been found more suitable to discover patterns for the minority class than conviction (Abdellatif, Ben Hassine, and Ben Yahia 2019). In the analysis of changes from older to more recent Teton Sioux songs, the different bias of the interest and conviction measures is reflected in the respective ratings of patterns such as ⟨compass : twelve_or_more_tones,old⟩ and ⟨firstNoteReKey : octave,modern⟩. Both mentioned patterns occur in around 30% of songs in the target class and just over 20% of songs in the background (see Table 3). In the first case, the target class old is the majority class (61%) in the corpus, while in the second case the target class modern is the minority class (39%). The majority-class pattern is ranked higher by conviction (level **A**) than by interest (level **B**), while the minority-class pattern is ranked higher by interest (level **A**) than by conviction (level **B**).

In terms of formal measure properties, interest is invariant to scaling the first column of the contingency table by a positive factor while conviction increases (Table 4, column M2.2). Scaling the first column increases the size of the target class, both in terms of its absolute count and also, with N remaining constant, in terms of its proportion in the corpus.

In summary, referring to the visual comparison of computational against reference interestingness ratings (Figure 3), these differences between the computational measures and their properties are reflected in high ranks (shown as dark blue cells) assigned to patterns at reference level **A** based on differences in relative frequencies between target class and background (dark blue cells for e.g. p-value, support difference, and PS); at reference level **B** biased towards infrequent contrast patterns (dark blue cells for growth rate, conviction, and interest); at reference level **C** biased towards patterns for the majority class (dark blue cells for confidence and IC^{++}); and at reference level **D** based on pattern frequency without contrasting target class and background (dark blue cells for coverage, support, and sensitivity).

4. Conclusions

In pattern mining, pattern interestingness measures distinguish potentially interesting from uninteresting patterns. In this paper we have presented a strategy and case study of exploiting interestingness measures and their properties to analyse patterns suggested by given music corpus studies. The contribution of this work is threefold.

First, computational pattern evaluation can support the meta-analysis of extant music corpus analyses. In the case of Densmore's comparison between old and modern Teton Sioux songs, the results of the computational analysis confirm Densmore's interest in differences between the two classes of songs – which is also reflected in the definition of the reference levels – complemented by a potential slight preference for more frequent patterns: "The first *important* point of *difference* is that the older songs show a much larger proportion having a range of 12 or more tones", while in "perhaps the *least important* of the tables [...] the groups show *no marked differences*" (Densmore 1918, 24, our emphasis). The concluding sentences of Densmore's summary refer to "*contrasts* between the two groups" (Densmore 1918, 25, our emphasis). Besides measures for contrasting patterns, including growth rate or support difference (Dong and Li 1999; Bay and Pazzani 2001), measures for dependent patterns, such as the PS measure and interest, have also been explicitly used for contrast mining, the latter in combination with a statistical significance test (Webb, Butler, and Newlands 2003; Novak et al. 2009).

Second, the method can be useful in analysing ground-truth, or reference, patterns employed for assessing pattern discovery algorithms. Quantitative studies of music pattern discovery using metrics such as precision and recall with respect to ground-truth patterns assess discovery algorithms equally on all patterns in the ground-truth set, thus implicitly assuming all ground-truth patterns to be of the same pattern type (e.g. van Kranenburg and Conklin 2016; Nuttall et al. 2019 for inter-opus pattern discovery in partitioned corpora; see also de Reuse and Fujinaga 2019). An analysis of the ground-truth patterns by computational interestingness measures can provide insights into the pattern types and the homogeneity or potential heterogeneity of the ground-truth pattern set. Existing studies on music pattern discovery algorithms have focused on sequential patterns; the pattern evaluation strategy presented in this paper can be easily extended from global-feature to sequential inter-opus patterns by defining a suitable pattern predicate (Conklin 2010a). A challenge, however, lies in the definition of the ground-truth patterns: if score annotations only identify patterns in the respective target class (e.g. van Kranenburg, Volk, and Wiering 2012), pattern interestingness measures which consider pattern counts in the background cannot be computed.

Third, insights from studying pattern interestingness measures and their behaviour in the context of extant music corpus analyses can inform inductive pattern discovery, more specifically the selection of pattern evaluation measures depending on, e.g. desired pattern types, expected or required pattern frequency, or the class distribution in the corpus. Regarding Densmore's studies, computational pattern mining also supports extending analysis beyond Densmore's single features to feature-set patterns (Neubarth, Shanahan, and Conklin 2018). The comparison of older and more recent Sioux songs focuses on distinctive patterns, which can be discovered by contrast, dependency, or significance measures, but other pattern types, such as characteristic or associative patterns, may equally be of interest (Densmore 1922, 1923).

Acknowledgments

The authors would like to thank the two anonymous reviewers and the journal editors for their valuable and helpful comments.

Disclosure statement

No potential conflict of interest was reported by the author(s).

ORCID

Darrell Conklin http://orcid.org/0000-0002-2313-9326

References

Abdellatif, Safa, Mohamed Ali Ben Hassine, and Sadok Ben Yahia. 2019. "Novel Interestingness Measures for Mining Significant Association Rules From Imbalanced Data." In *Web, Artifical Intelligence and*

Network Applications (WAINA 2019), edited by Leonard Barolli, Makoto Takizawa, Fatos Xhafa, and Tomoya Enokido, 172–182. Cham: Springer.

Bay, Stephen D., and Michael J. Pazzani. 2001. "Detecting Group Differences: Mining Contrast Sets." *Data Mining and Knowledge Discovery* 5 (3): 213–246.

Brin, Sergey, Rajeev Motwani, and Craig Silverstein. 1997. "Beyond Market Baskets: Generalizing Association Rules to Correlations." *ACM SIGMOD Record* 26 (2): 265–276.

Brin, Sergey, Rajeev Motwani, Jeffrey D. Ullman, and Shalom Tsur. 1997. "Dynamic Itemset Counting and Implication Rules for Market Basket Data." In *Proceedings of the ACM SIGMOD International Conference on Management of Data (SIGMOD'97)*, Tucson, Arizona, USA, 255–264.

Broze, Yuri, and Daniel Shanahan. 2013. "Diachronic Changes in Jazz Harmony: A Cognitive Perspective." *Music Perception* 31 (1): 32–45.

Collins, Tom, Andreas Arzt, Harald Frostel, and Gerhard Widmer. 2016. "Using Geometric Symbolic Fingerprinting to Discover Distinctive Patterns in Polyphonic Music." In *Computational Music Analysis*, edited by David Meredith, 445–474. Cham: Springer.

Conklin, Darrell. 2010a. "Discovery of Distinctive Patterns in Music." *Intelligent Data Analysis* 14, 547–554.

Conklin, Darrell. 2010b. "Distinctive Patterns in the First Movement of Brahms' String Quartet in C Minor." *Journal of Mathematics and Music* 4 (2): 85–92.

Conklin, Darrell. 2013. "Antipattern Discovery in Folk Tunes." *Journal of New Music Research* 42 (2): 161–169.

Conklin, Darrell, and Christina Anagnostopoulou. 2001. "Representation and Discovery of Multiple Viewpoint Patterns." In *Proceedings of the International Computer Music Conference (ICMC 2001)*, Havana, Cuba, 479–485.

Conklin, Darrell, and Christina Anagnostopoulou. 2011. "Comparative Pattern Analysis of Cretan Folk Songs." *Journal of New Music Research* 40 (2): 119–125.

de Reuse, Timothy, and Ichiro Fujinaga. 2019. "Pattern Clustering in Monophonic Music by Learning a Non-Linear Embedding from Human Annotations." In *Proceedings of the 20th International Society for Music Information Retrieval Conference (ISMIR 2019)*, Delft, Netherlands, 761–768.

Densmore, Frances. 1910. *Chippewa Music*. Washington, DC: Smithsonian Institution, Bureau of American Ethnology, Bulletin 45.

Densmore, Frances. 1918. *Teton Sioux Music*. Washington, DC: Smithsonian Institution, Bureau of American Ethnology, Bulletin 61.

Densmore, Frances. 1922. *Northern Ute Music*. Washington, DC: Smithsonian Institution, Bureau of American Ethnology, Bulletin 75.

Densmore, Frances. 1923. *Mandan and Hidatsa Music*. Washington, DC: Smithsonian Institution, Bureau of American Ethnology, Bulletin 80.

Dong, Guozhu, and Jinyan Li. 1999. "Efficient Mining of Emerging Patterns: Discovering Trends and Differences." In *Proceedings of the 5th ACM SIGKDD International Conference on Knowledge Discovery and Data Mining (KDD-99)*, San Diego, CA, USA, 43–52.

Geng, Liqiang, and Howard J. Hamilton. 2006. "Interestingness Measures for Data Mining: A Survey." *ACM Computing Surveys* 38 (3): 1–32.

Han, Jiawei, Hong Cheng, Dong Xin, and Xifeng Yan. 2007. "Frequent Pattern Mining: Current Status and Future Directions." *Data Mining and Knowledge Discovery* 15, 55–86.

Han, Jiawei, Yongjian Fu, Wei Wang, Krzysztof Koperski, and Osmar Zaiane. 1996. "DMQL: A Data Mining Query Language for Relational Databases." In *Proceedings of the ACM SIGMOD Workshop on Data Mining and Knowledge Discovery (DMKD-96)*, Montreal, Canada, 27–33.

Jackson, Roland. 1970. "Harmony Before and After 1910: A Computer Comparison." In *The Computer and Music*, edited by Harry B. Lincoln, 132–146. Ithaca, NY: Cornell University Press.

Kamber, Micheline, and Raijan Shinghal. 1996. "Evaluating the Interestingness of Characteristic Rules." In *Proceedings of the 2nd International Conference on Knowledge Discovery and Data Mining (KDD-96)*, Portland, OR, USA, 263–266.

Kavšek, Branko, and Nada Lavrač. 2006. "APRIORI-SD: Adapting Association Rule Learning to Subgroup Discovery." *Applied Artificial Intelligence* 20, 543–583.

Lavrač, Nada, Peter Flach, and Blaz Zupan. 1999. "Rule Evaluation Measures: A Unifying View." In *Proceedings of the 9th International Workshop on Inductive Logic Programming (ILP-99)*, Bled, Slovenia, 174–185.

Lenca, Philippe, Benoît Vaillant, Patrick Meyer, and Stéphane Lallich. 2007. "Association Rule Interestingness Measures: Experimental and Theoretical Studies." In *Quality Measures in Data Mining*, edited by Fabrice Guillet and Howard J. Hamilton, 51–76. Berlin, Heidelberg: Springer.

Li, Jundong, and Osmar R. Zaiane. 2017. "Exploiting Statistically Significant Dependent Rules for Associative Classification." *Intelligent Data Analysis* 21 (5): 1155–1172.

Liu, Bing, Wynne Hsu, and Yiming Ma. 1998. "Integrating Classification and Association Rule Mining." In *Proceedings of the 4th International Conference on Knowledge Discovery and Data Mining (KDD-98)*, New York, USA, 80–86.

Neubarth, Kerstin, Daniel Shanahan, and Darrell Conklin. 2018. "Supervised Descriptive Pattern Discovery in Native American Music." *Journal of New Music Research* 47 (1): 1–16.

Novak, Petra Kralj, Nada Lavrač, Dragan Gamberger, and Antonija Krstačić. 2009. "CSM-SD: Methodology for Contrast Set Mining Through Subgroup Discovery." *Journal of Biomedical Informatics* 42 (1): 113–122.

Nuttall, Thomas, Miguel García Casado, Víctor Núñez Tarifa, Rafael Caro Repetto, and Xavier Serra. 2019. "Contributing to New Musicological Theories with Computational Methods: The Case of Centonization in Arab-Andalusian Music." In *Proceedings of the 20th International Society for Music Information Retrieval Conference (ISMIR 2019)*, Delft, Netherlands, 223–228.

Ohsaki, Miho, Shinya Kitaguchi, Kazuya Okamoto, Hideto Yokoi, and Takahira Yamaguchi. 2004. "Evaluation of Rule Interestingness Measures with a Clinical Dataset on Hepatitis." In *Proceedings of the 8th European Conference on Principles and Practice of Knowledge Discovery in Databases (PKDD 2004)*, Pisa, Italy, 362–373.

Piatetsky-Shapiro, Gregory. 1991. "Discovery, Analysis and Presentation of Strong Rules." In *Knowledge Discovery in Databases*, edited by Gregory Piatetsky-Shapiro and William Frawley, 229–248. Cambridge, MA: MIT Press.

Shanahan, Daniel, Kerstin Neubarth, and Darrell Conklin. 2016. "Mining Musical Traits of Social Functions in Native American Music." In *Proceedings of the 17th International Society for Music Information Retrieval Conference (ISMIR 2016)*, New York, USA, 681–687.

Taminau, Jonatan, Ruben Hillewaere, Stijn Meganck, Darrell Conklin, Ann Nowé, and Bernard Manderick. 2009. "Descriptive Subgroup Mining of Folk Music." In *2nd International Workshop on Machine Learning and Music at ECML/PKDD 2009 (MML 2009)*, Bled, Slovenia.

Tan, Pang-Ning, Vipin Kumar, and Jaideep Srivastava. 2002. "Selecting the Right Interestingness Measure for Association Patterns." In *Proceedings of the 8th ACM SIGKDD International Conference on Knowledge Discovery and Data Mining (KDD 2002)*, Edmonton, AB, Canada, 32–41.

van Kranenburg, Peter, and Darrell Conklin. 2016. "A Pattern Mining Approach to Study a Collection of Dutch Folk Songs." In *6th International Workshop on Folk Music Analysis (FMA 2016)*, Dublin, Ireland.

van Kranenburg, Peter, Anja Volk, and Frans Wiering. 2012. "On Identifying Folk Song Melodies Employing Recurring Motifs." In *Proceedings of the 12th International Conference on Music Perception and Cognition and the 8th Triennial Conference of the European Society for Cognitive Sciences of Music (ICMPC/ESCOM 2012)*, Thessaloniki, Greece, 1057–1062.

Webb, Geoffrey I. 2007. "Discovering Significant Patterns." *Machine Learning* 68, 1–33.

Webb, Geoffrey I., Shane Butler, and Douglas Newlands. 2003. "On Detecting Differences Between Groups." In *Proceedings of the 9th ACM SIGKDD International Conference on Knowledge Discovery and Data Mining (KDD-2003)*, Washington, DC, USA, 256–265.

A computational exploration of melodic patterns in Arab-Andalusian music

Thomas Nuttall, Miguel G. Casado, Andres Ferraro, Darrell Conklin, and Rafael Caro Repetto

Here we present a computational approach to identifying melodic patterns in a dataset of 145 MusicXML scores with the aim of contributing to centonization theory in the Moroccan tradition of Arab-Andalusian Music – a theory in development by expert performer and researcher of this tradition, Amin Chaachoo. Central to his work is the definition of a set of characteristic patterns, or centos, for each *ṭab'*, or melodic mode. We apply three methods: TF-IDF, Maximally General Distinctive Patterns (MGDP) and the Structure Induction Algorithm (SIA) to identify characteristic patterns at the level of *ṭab'*. A substantial number of the centos proposed by Chaachoo are identified and new melodic patterns are retrieved. A discussion with Chaachoo about the obtained results promoted the elicitation of other categories of recurrent patterns in the tradition different from the centos, contributing to a deeper musicological knowledge of the tradition.

1. Introduction

1.1. *Characteristics of Arab-Andalusian music*

Arab-Andalusian music formed in the medieval Islamic territories of the Iberian Peninsula, known as Al-Andalus, as a result of the combination of local musical traditions with Arab poetry and aesthetics from the Middle East. Its core element is the *ṣan'a* (plural *ṣanā'i'*), a poem sung by a choir accompanied by an instrumental ensemble. These *ṣanā'i'* are performed in suites known as *nawabāt* (plural of *nawba*), which include orchestral pieces and both instrumental and vocal solo improvisations (Guettat 2000; Chaachoo 2016). The *nawba* is the essential form of Arab-Andalusian music. Traditionally, all *ṣanā'i'* and other pieces in one *nawba* are composed in one single melodic mode, known as *ṭab'* (plural *ṭūbu'*).

With the migration of the Andalusian population to North Africa following the defeat of Islamic kingdoms in the Iberian peninsula, Arab-Andalusian music was taken to North Africa, where it has been preserved until today. Nowadays, it is the classical music repertoire in countries

Figure 1. First measures of the first *ṣan'a* in the score *Bṭayḥī al-ḥiǧāz al-kabīr* (with the MBID 12ce112f-38ed-4700-94ec-a329d06f6196). Boxes with the same border type indicate occurrences of the same *cento*.[1]

such as Morocco, Algeria and Tunisia, in each of which it developed into a particular tradition (Poché 1997; Guettat 2000). In this paper, we focus on the Moroccan repertoire, known as *al-Āla* (Chaachoo 2016).

Over the past decade, the researcher and expert performer of *al-Āla* Amin Chaachoo has been developing a music theory for this tradition (Chaachoo 2011, 2016, 2019). One of the most relevant hypotheses of Chaachoo's theory is the Iberian roots of the melodic dimension of Arab-Andalusian music, specifically arguing that the *ṭab'* exhibits different characteristics to Middle Eastern *maqam* (Chaachoo 2019, 19). According to this claim, Chaachoo explains *ṭūbu'* in terms of the modal theory developed for plainchant. Thus, a *ṭab'* is defined by a particular ascending and descending scale, a fundamental degree similar in function to the *finalis* of Gregorian modes, several principal degrees, one or two persistent degrees in the manner of "the reciting tone" and a series of characteristic melodic phrases.

As a theoretical framework for the function of these characteristic melodic phrases in Arab-Andalusian *ṭūbu'*, Chaachoo draws on the *centonization* model. Centonization, from Latin *cento* meaning patchwork, is defined as a plainchant compositional technique consisting of the combination of pre-existing melodic units called *centos* (Ferretti and Agaësse 1938; Apel 1958; Chewand and McKinnon 2001). Chaachoo argues that melodies in Arab-Andalusian music are also created by the combination of *centos*, thus strengthening the connection between this tradition and Iberian local practices. Consequently, the identity of a particular *ṭab'* is also defined by a corresponding set of such *centos*, see Figure 1. In each new publication of his theory, Chaachoo presents a slightly modified list of *centos* per *ṭab'*.

1.2. *Motivation and objectives*

Chaachoo's centonization theory for *al-Āla* is in continuous development and we have available to us a transcribed symbolic *ṭab'*-annotated collection of many hours of performance in this tradition. We attempt to contribute to this developing theory with an empirical investigation into the extent to which Chaachoo's *centos* characterize the melodic content of each *ṭab'*, further suggesting other salient melodic patterns identified by our approach as being characteristic.

It is important to note that the *centos* proposed by Chaachoo cannot be considered a strict ground truth and as such the main objective of this paper is not to validate our pattern discovery approaches on this collection but instead to explore whether we can validate parts of Chaachoo's theory empirically, suggesting new and interesting areas of investigation for its continual development.

[1] The full score is available in *MuseScore* (https://musescore.com/mtg/brihiorchestra_rtm1960s_btayhihijazkabir).

1.3. *Previous work*

Automatic symbolic pattern discovery is an active research area in Music Information Retrieval that aims to automatically identify *musically meaningful* patterns in symbolic representations of music such as scores. An example of this is the MIREX challenge, *Discovery of Repeated Themes & Sections*, proposed until 2017.[2]

There exists many studies into melodic pattern discovery in symbolic scores, summaries of which have been made by Janssen et al. (2013) and more recently by Ren et al. (2017). Lack of agreement on the current state-of-the-art stems from the difficulty in evaluating approaches, with expertly annotated ground truth often required for performance measurement, more often than not on a study-by-study basis.

The task of pattern discovery within the context of Arab-Andalusian music is a much more recent development, with Nuttall et al. (2019) taking advantage of the Arab-Andalusian corpus gathered as part of the CompMusic project to identify patterns characteristic of *nawabāt*. Since then, Chaachoo has published a new and revised version of his centonization theory (Chaachoo 2019), offering us the opportunity to explore this tradition from a new perspective with a wider range of methods.

2. Dataset

Our dataset is a subset of the Music Scores Collection of the Arab-Andalusian Music Corpus gathered in the context of the CompMusic project (Serra 2014).[3] This corpus is the largest source of machine readable data for the computational study of this music tradition.

The Arab-Andalusian Music Corpus comprises three different but related collections: (1) The Audio Recordings Collection – containing 156 audio recordings from the personal collection of Amin Chaachoo, donated to the CompMusic project by him and which are now available under Creative Commons license. These recordings have a mean duration of one hour and contain performances on radio and at private events from the 1960s and 1970s. They are selected for the performance quality of the featured orchestras, namely the Tetouan Orchestra, Orchestra of the Conservatory of Tetouan, Brihi Orchestra and RTM Orchestra, which include celebrated maestros of *al-Āla*, (2) The Music Scores Collection consists of 158 manual transcriptions of audio recordings done by Amin Chaachoo. Since Arab-Andalusian music is heterophonic in texture, only the predominant melody is transcribed, which underlies the specific renditions by the different instruments in the orchestra and by the choir (see Figure 1). Solo improvisations have not been transcribed. The scores are stored as MusicXML files and are also available under Creative Common license, (3) The Lyrics Collection – containing non-aligned lyrics of all recordings, both in their original Arabic script and in an automatically generated romanization (Sordo, Chaachoo, and Serra 2014). The metadata of the recordings is stored in MusicBrainz – the scores, audio and lyrics of an individual performance are connected by and stored under a common MusicBrainz ID (MBID).[4]

The corpus is integrated into Dunya (Porter, Sordo, and Serra 2013), from where it can be retrieved using a series of Jupyter Notebooks.[5] It can be also downloaded in bulk from a Zenodo repository.[6]

---

[2] https://www.music-ir.org/mirex/wiki/2017:Discovery_of_Repeated_Themes_%26_Sections
[3] https://compmusic.upf.edu/
[4] https://musicbrainz.org/collection/142ea0d7-7fdf-4ea5-9b04-219f68023d01
[5] https://github.com/MTG/andalusian-corpus-notebooks
[6] https://doi.org/10.5281/zenodo.1291775

Table 1. Distribution of scores across *ṭūbu'*.

ṭab' name	Number of scores
al-istihlāl	23
al-iṣbahān	13
al-māšriqī	10
al-māya	12
al-raṣd	10
al-ḥiŷāz al-kabīr	10
al-ḥiŷāz al-māšriqī	5
al-sīka	1
al-'uššāq	7
garībat al-ḥusayn	12
raml al-māya	19
raṣd al-ḏāyl	16
'irāq al-'aŷam	7

Our dataset consists of 145 scores from the Music Scores Collection of the Arab-Andalusian music corpus, having discarded the scores from the collection that do not belong to any of the classical *ṭūbu'*. The metadata of the scores – inherited from the recordings – specify the *nawba* and rhythmic mode of each piece. The scores have been manually segmented in terms of structural sections according to the manual segmentation of the recordings done by Chaachoo and also inherit his annotations of form and *ṭab'*. According to these annotations, the corpus covers 13 of the 26 classical *ṭūbu'* theoretically established for the *al-Āla* tradition. The distribution of scores across *al-Āla* is shown in Table 1.

3. Methodology

We apply three pattern discovery techniques to our symbolic dataset in an attempt to identify which melodic patterns are most characteristic of each *ṭab'*, cross-referencing our findings with Chaachoo's (2019) latest publication on centonization theory.

3.1. *TF-IDF*

Our first method is an approach first introduced by Nuttall et al. (2019) on the same dataset in which characteristic patterns are investigated at the level of *nawba*. Here we use the same methodology to identify representative patterns on a *ṭab'* level.

3.1.1. *Data representation*

Each score is represented by a "bag of patterns" (analogous to a bag-of-words representation in natural language processing) in which *n-grams* – in this instance concatenated sequences of consecutive notes – up to a pre-specified length, N, are extracted from each score to build an unordered multi-set of patterns that exist within it. Any n-gram that contains a rest, R, is discarded, as is any temporal information. For a score of $[G, E, F, F, R, E, G, E]$, the bag-of-patterns representation for $N = 2$, 3 and 4 is as follows:

$\mathrm{N} = 2$: [GE, EF, FF, EG, GE];
$\mathrm{N} = 3$: [GE, GEF, EF, EFF, FF, EG, EGE, GE];
$\mathrm{N} \geq 4$: [GE, GEF, GEFF, EF, EFF, FF, EG, EGE, GE]

3.1.2. *TF-IDF statistic*

For each pattern, p, and each score, s, a TF-IDF statistic with sub-linear tf scaling is computed to measure the extent to which each pattern is *over/under-represented* in the score. The *term-frequency, inverse document frequency* and TF-IDF statistic are presented in equations (1), (2) and (3), respectively. $df(p)$ is the raw count of scores in which pattern p occurs; $f_{p,s}$ is the raw count of occurrences of pattern p in score s; n is the total number of scores.

$$\mathrm{tf}(p, s) = 1 + log(f_{p,s}) \tag{1}$$

$$\mathrm{idf}(p) = log\left(\frac{1+n}{1+\mathrm{df}(p)}\right) + 1 \tag{2}$$

$$\text{TF-IDF}(p, s) = \mathrm{tf}(p, s) \cdot \mathrm{idf}(p) \tag{3}$$

Patterns that occur more frequently than would be expected for a given score will return a higher TF-IDF statistic and are considered *characteristic* of the score.

With the TF-IDF statistic as a measure of how characteristic each pattern is of each score we group scores by our *ṭab‘* annotations and average the statistic for each pattern (as in Nuttall et al. 2019). The result is a measure of how characteristic each pattern is of each *ṭab‘*.

3.2. ***Structure induction algorithm***

Structure Induction Algorithm (SIA) (Meredith, Lemström, and Wiggins 2002; Meredith and Wiggins 2001) is one of the most popular methods for symbolic pattern discovery, basing its approach on geometric methods. The algorithm has recently been applied to Western music with successful results on a variety of tasks such as genre classification (Ferraro and Lemström 2018) and melodic pattern detection in Dutch folk music (Janssen, van Kranenburg, and Volk 2015). We are not interested in translations of patterns, i.e. we do not consider two instances of the same progression in different keys as equal. For this reason, we choose to use the SIA algorithm rather than the more specialised SIATEC.

3.2.1. *Discovering maximal repeated patterns in multidimensional datasets*

SIA represents scores as data collections, D, of two dimensions: onset time and pitch. Each score is made up of notes $d_1, d_2, \ldots d_n$. A pattern P is translatable by a vector v in a data collection D if and only if P can be translated by v to give a pattern that is a subset of D. Fundamental to SIA is the concept of the *maximal translatable pattern (MTP)*. Formally, the *MTP* for a vector v in a data collection D, denoted by *MTP(v, D)*, is the largest pattern translatable by v in D. Mathematically:

$$MTP(v, D) = \{d \mid d \in D \wedge d + v \in D\}. \tag{4}$$

In music, *MTPs* often correspond to the patterns involved in perceptually significant repetitions. The main goal of SIA is to compute all non-empty *MTPs* in a data collection. The *MTP* for a vector, v, in a data collection, D, is defined as non-empty if and only if there exists at least two data points d_1 and d_2 in D such that $v = d_2 - d_1$. To sum up, SIA finds for every possible vector the largest pattern in the data collection that can be translated by that vector to give another pattern in the data collection (Meredith, Lemström, and Wiggins 2002).

SIA is applied independently to each score of the dataset and the results grouped by *ṭab‘* to generate a list of relevant patterns on a *ṭab‘* level. Patterns that consist of notes that are not consecutive in the score are discarded so as to make the results comparable with the other algorithms used in this paper that do not have the ability of finding patterns of non-consecutive notes.

3.3. *Maximally general distinctive patterns (MGDP)*

Maximally general distinctive pattern mining (MGDP) (Conklin 2010) is a method designed to find general patterns that are also distinctive. A *distinctive pattern* is one that is frequent and over-represented in a positive corpus (here, scores in a particular *ṭab‘*) as compared to an anticorpus (all other scores). The degree of over-representation of a pattern Φ is measured by its relative frequency between the positive corpus (denoted $\oplus$) and anticorpus (denoted $\ominus$):

$$\Delta = \frac{P(\Phi \mid \oplus)}{P(\Phi \mid \ominus)}$$

In the above, $P(\Phi \mid \oplus)$ and $P(\Phi \mid \ominus)$ are the relative frequencies of the pattern Φ in the positive corpus and anticorpus, respectively.

A *maximally general* distinctive pattern is a pattern with $\Delta \geq \epsilon$ and for which no containing pattern is also distinctive. In this paper, $\epsilon = 3$ for distinctiveness is used, signifying a three-fold or higher over-representation in the positive corpus. To find maximally general distinctive patterns, a containment hierarchy of patterns is traversed from general to specific, terminating the search at a branch when a distinctive pattern is encountered.

3.4. *Pattern selection*

To take all patterns returned by any of our three methods would result in an unmanageable quantity of characteristic patterns. Two pattern selection processes are applied in each approach.

3.4.1. *Minimum frequency of occurrence threshold (MFO)*

As noted by Nuttall et al. (2019), many longer, rarer patterns (that occur infrequently across the whole dataset) have a disproportionately high *importance* as determined by each of our models, this in reality is a reflection of infrequency of occurrence, rather than correlation between occurrence and *ṭab‘*. We consider the infrequency of such patterns as not providing us with enough *signal* to have confidence in this *importance* and like Nuttall et al. (2019), impose a *minimum frequency of occurrence* (MFO) for each pattern per score per *ṭab‘*.

Our *minimum frequency of occurrence threshold* of 59 per *ṭab‘* dictates that for a particular *ṭab‘*, a pattern is discarded if the number of occurrences of that pattern in scores annotated with that *ṭab‘* divided by the number of scores annotated with that *ṭab‘* is less than the minimum frequency of occurrence, 59. This threshold is applied to all three methods and determined by maximizing the F_1 score (harmonic mean of recall and precision) when evaluated on the *centos* provided by Chaachoo (2019). It is important to note that although the *centos* provided by Chaachoo are not necessarily a canonical ground truth, they are the most useful guide we have in parameter selection and that many of Chaachoo's patterns themselves do not reach this minimum frequency. Indeed some patterns proposed by him do not exist at all in the dataset (more on this in Section 5). It is also true that by imposing such a threshold we limit our output to patterns of shorter lengths (since longer ones are less common), this is understood and considered a necessary limitation of the pattern-space we want to explore.

3.4.2. *Pattern character*

The *centos* as outlined by Chaachoo (2019) are monophonic sequences of notes, each with at least two unique pitches, without duration or octave and length ranging between and including 3

Table 2. Evaluation on Chaachoo's patterns above minimum frequency occurrence.

ṭab'	Recall	Precision	# Centos	# Retrieved
al-istihlāl	1.00	0.26	6	23
al-iṣbahān	0.83	0.33	6	15
al-māšriqī	0.75	0.30	4	10
al-ḥiŷāz al-māšriqī	n/a	0.00	0	15
al-māya	1.00	0.21	3	14
al-raṣd	1.00	0.25	4	16
al-ḥiŷāz al-kabīr	1.00	0.19	3	16
al-sīka	1.00	0.25	1	4
al-'uššāq	1.00	0.22	2	9
garībat al-ḥusayn	1.00	0.25	4	16
raml al-māya	0.75	0.23	4	13
raṣd al-ḏāyl	1.00	0.17	3	18
'irāq al-'aŷam	1.00	0.29	5	17
total	**0.93**	**0.25**	**45**	**186**

and 7 notes. As such we limit our search to operate within this space by

- Considering only patterns between and including lengths of 3 and 7 notes
- Removing duration and octave from the data representation
- Not considering as a pattern any sequence of notes that includes a rest
- Discarding "patterns" that consist of only one unique pitch

The code to explore the dataset using some of our approaches can be found in the project repository on Github.[7]

4. Results

Each model described in Section 3 is applied to our dataset. The full output of each can be found in the results directory of the accompanying Github repository.

Table 2 displays the recall and precision of our three methods combined on a *ṭab'* level, **# Centos** indicate how many of Chaachoo's *centos* were available to find and **# Retrieved** is the number of unique patterns returned by all models. We consider Chaachoo's *centos* that occur more frequently than our *minimum frequency of occurrence* as ground truth. Also displayed are the number of Chaachoo's *centos* above this threshold for each *ṭab'* and how many in total were discovered by our methods.

In total, we find 186 significant patterns across all *ṭūbu'*, 42 of which correspond to the 45 of Chaachoo's above the MFO threshold. Resulting in a total recall of 0.93 and total precision of 0.25. A full breakdown of results can be found in the Github repository.

4.1. *Distinctive patterns*

One of the interesting features of the MGDP approach (Section 3.3) is that it offers us an intuitive comparison of a pattern's relative probability within a *ṭab'* and outside of a *ṭab'* (distinctiveness Δ, Section 3.3). Table 3 presents these deltas for Chaachoo's patterns to reflect their ability to be recalled when mining for all patterns subject to our selection processes in Section 3.4. The table is sorted from high to low distinctiveness, Δ. For every pattern, its global rank (the position

[7] https://github.com/centonization/centonizationtheory/

Table 3. Distinctiveness of Chaachoo's *centos*.

			rank	
ṭab'	Pattern	Δ	Global	Local
al-ḥiŷāz al-kabīr	AGF#E-D	62.58	5	5
'irāq al-'aŷam	GF#ED	13.14	11	3
al-ḥiŷāz al-kabīr	F#GA	7.25	20	11
'irāq al-'aŷam	EF#G	6.75	21	5
al-raṣd	F#GA	5.52	25	7
al-māya	B-AG	4.01	35	3
al-raṣd	AGE	3.99	36	9
'irāq al-'aŷam	BAG	2.71	46	9
al-iṣbahān	BCD	2.57	51	4
raṣd al-ḏāyl	CDE	2.31	55	5
al-ḥiŷāz al-kabīr	CBAG	2.31	57	13
al-istihlāl	FAG	2.18	66	2
al-istihlāl	ABC	2.09	71	3
al-'uššāq	BAG	2.02	75	1
al-sīka	AGFE	1.89	82	2
al-māšriqī	FGA	1.88	84	2
raml al-māya	FGA	1.84	88	1
al-māya	EFG	1.78	96	6
raṣd al-ḏāyl	EDC	1.57	111	16
garībat al-ḥusayn	CDE	1.57	112	3
al-istihlāl	FEDC	1.52	118	8
al-iṣbahān	FEDC	1.48	125	12
al-istihlāl	GFE	1.46	127	10
al-iṣbahān	FED	1.46	128	13
raṣd al-ḏāyl	AGFE	1.45	131	20
al-raṣd	CDE	1.44	134	18
al-iṣbahān	EFG	1.40	141	16
al-māya	AGFE	1.36	150	8
garībat al-ḥusayn	AGFE	1.34	154	7
al-istihlāl	CBAG	1.29	163	20
al-istihlāl	EFG	1.20	181	22
al-māšriqī	AGF	1.19	182	4
garībat al-ḥusayn	FED	1.16	190	10
al-iṣbahān	GFE	1.13	194	20
garībat al-ḥusayn	AGF	1.12	198	12
raml al-māya	EFG	1.03	215	13
raml al-māya	FED	1.00	218	14
al-māšriqī	FED	0.98	223	7
al-raṣd	EDC	0.96	225	19
al-iṣbahān	AGFE	0.93	230	22
al-'uššāq	FED	0.87	236	9
raml al-māya	FEDC	0.84	239	15
al-māšriqī	FEDC	0.80	241	11
'irāq al-'aŷam	EDC	0.58	254	17
'irāq al-'aŷam	FED	0.48	258	19

when all patterns for all classes are concatenated together) and its local rank (position within the patterns for the particular target) are indicated. These results indicate that of the 45 *centos* occurring above the minimum frequency threshold, 8 (at the bottom of Table 3) have a Δ below 1.00 which indicates that they are in fact under-represented in the particular *ṭab'*.

5. Discussion

The exploratory study described in this paper and particularly the obtained results presented in the previous section produce valuable contributions to the deeper understanding of the analysed dataset, the selected methods and the very task itself from both a musicological and technical perspective. This work has also given us the opportunity to contact Chaachoo, who shared with us his thoughts on the research carried out and its output. It is beyond the scope of this paper to engage in a detailed discussion of the obtained results for each *ṭab‘*. However, we will highlight in the current section the major outcomes of this process.

In previous studies (Sordo, Chaachoo, and Serra 2014; Caro Repetto et al. 2018), the Music Scores Collection of the Arab-Andalusian Music Corpus has been analysed according to the five criteria proposed by Serra (2014). Here we focus on its coverage for the specific task of melodic pattern analysis. Table 1 shows that the collection covers 13 of the 26 classical *ṭūbu‘* and not all of them equally (since the distribution of scores across *ṭūbu‘* is uneven). Most remarkable is the case of the *ṭab‘ al-sīka*, for which only one score is available, an exceptionally short one, corresponding to a recording of 7’13”. Consequently, only one of the 6 *centos* proposed by Chaachoo for this *ṭab‘* is found above the MFO (see Section 3.4), and two of them are not present at all in the available score. Also worth noting is the case of *al-ḥiŷāz al-māšriqī*, for which 5 scores are available (audio duration of 3h2’25”), but none of the 3 *centos* proposed by Chaachoo are present above the MFO. Despite the direct implications that this has for our applied methods, there are still interesting outcomes. In the case of *al-sīka* both TF-IDF and SIA found the only *cento* which occurs above the MFO, AGFE. This pattern has a local rank in terms of delta of 2 (Table 3), and it ranks 3rd in TF-IDF and 2nd in SIA. This result hence contributes to consolidate its significance for this *ṭab‘*, as confirmed by Chaachoo.

A fundamental contribution of using computational methods for musicological research is the opportunity to make implicit knowledge from musicians or expert *connoisseurs* of the studied music tradition explicit. This is precisely the case for our study, whose discussion with Chaachoo highlighted knowledge that was only briefly mentioned in his publications (if at all) but that is essential for the goals addressed in our research.

Although Chaachoo acknowledges the usefulness of computational analysis in the revision of his theory, it was apparent from our discussion that his main source for evaluation is his own experiential knowledge and, as he stated, “intuition .” Therefore, in order to evaluate the importance of the patterns retrieved from the algorithms, his only criterion was his experience with the music – the associated metrics reported by the used methods did not contribute to his evaluation. The inspection of the retrieved patterns gave Chaachoo the opportunity to specify the existence of melodic patterns in the *al-Āla* tradition that are not considered as defining of the *ṭūbu‘*, that is, that can not be considered *centos*, and therefore are not theorized in his publications. If *centos* are “minimal recurrent units” used as “building blocks” of Arab-Andalusian melodies, *melodic formulas* are recurrent combinations of *centos*. These are mentioned in Chaachoo’s publications only briefly and not studied in detail (2016, 221-22). The melodies of the specific pieces, besides *centos* and *melodic formulas* also contain piece specific material (see for example the melody not contained in boxes in Figure 1). Chaachoo observed that some of the retrieved patterns correspond to joining points of the two *centos* of a *melodic formula*, that is, the last note(s) and the first one(s) of two frequently joined *centos*, or the beginning or ending of a *cento* together with a note that commonly precedes or follows that *cento*. Besides, some of the retrieved patterns correspond, according to Chaachoo, to ornamentation preferences of the particular performing orchestra, a type of pattern not specified in his publications. And finally, Chaachoo also recognized some patterns that he would describe as “expressive trends” related to the tradition as a whole. The existence of such *non-cento* patterns was unknown by the authors, and their elicitation and definition is an outcome of the discussion precipitated by the study presented here.

In order to gain a more concrete understanding of Chaachoo’s evaluation of our results, we discussed the case of *al-ḥiŷāz al-kabīr* (Figure 1 offers a short example). Although 10 scores are available for *al-ḥiŷāz al-kabīr* (audio duration of 5h16’33”), only three out of the five *centos* from

Chaachoo occur over the MFO. Among them, the pattern AGF#E-D is the most distinctive of Chachoo's *centos* as shown in Table 3. The distinctiveness of this pattern is explained by the fact that it is one of the few that contains what Chaachoo defines as the *al-Zayadan* genre, that is, an intervalic sequence of minor second, augmented second and minor second. The very name of the *ṭab'*, literally "the great *ḥiŷāz*," suggests the relevance of this sequence, since *ḥiŷāz* is the name of a relevant Arabic *maqam*, also characterised by this sequence. This *al-Zayadan* genre confers to these *ṭūbu'* their very characteristic "oriental flavour ." However, Chaachoo also discusses how the importance of this sequence has been overemphasized recently and extensively used in improvised solos, but in fact does not appear in traditional instrumental preludes and many *ṣanā'i'* (2019, 257). Considering this last observation, we asked Chaachoo if the fact that only one of the tested methods retrieved the *cento* as it is proposed by him, but that many substrings of it are retrieved by the three methods around the characteristic augmented second might invite him to reconsider the contour of this *cento*. He maintains that he is certain of the relevance of this *cento* in its current contour and proposes possible reasons for it not being retrieved with higher relevance by our methods.

The discussion of the results related to *al-ḥiŷāz al-kabīr* resulted in a broader discussion on the concept and function of *centos* themselves. None of the results obtained in our study, even when a particular *cento* proposed by Chaachoo is not retrieved by any of the methods, is sufficient to disprove their significance. First of all, our dataset does not exhaustively cover the reality of the Arab-Andalusian music tradition, as does the knowledge upon which Chaachoo's theory is built. Beyond the coverage of our dataset, the contained music scores represent, as confirmed by Chaachoo, an intermediate level between the performative, sonic surface and the compositional, theoretical melodic skeleton, which is the deep level on which *centos* operate. In the intermediate level represented in the scores, *centos* might be "hidden ," as Chaachoo put it, behind an ornament produced by the aesthetic preferences of an orchestra or a more general expressive trend. This might point to a discrepancy between practice and theory, a fact that is known and commonly studied in other music traditions, such as in Turkish *makam* music (Bozkurt, Ayangil, and Holzapfel 2014) or in South Indian *rāga* music (Pearson 2016). Secondly, it should be considered that *centos* are part of a broader system, the *ṭab'*, and therefore, their relevance might arise from interaction with other elements of this system. In our discussion, Chaachoo emphasized how *centos* are directed to stress one of the relevant degrees of the *ṭab'*. For example, the above discussed *cento* for *al-ḥiŷāz al-kabīr*, AGF#E-D, not only includes the characteristic *al-Zayadan* genre, but also represents a movement from the persistent degree of the *ṭab'*, A, to its fundamental degree, D. Equally, the two *centos* below the minimum frequency threshold, CED and FED (see Figure 1 for the latter), are melodic movements towards the fundamental degree, a fact that reinforces their significance. Furthermore, looking at the context even more broadly, a melodic pattern might not only be relevant because of its statistical occurrence, but because of its structural location as pointed out by Volk and van Kranenburg (2001). Asked about this issue in Arab-Andalusian music, Chaachoo confirmed that *centos* tend to be used to conclude poetic/melodic phrases, and some of them are very characteristic for their use in concluding *ṣanā'i'*, as is the case of the discussed *cento* AGF#E-D of *al-ḥiŷāz al-kabīr*. All these reflections offer very important contributions for improving the automatic analysis of Arab-Andalusian *centos* in future work.

It is also worth reflecting on the three methods used in our study, inspecting the differences in patterns retrieved by each of them for *al-ḥiŷāz al-kabīr* (Table 4). Regarding Chaachoo's *cento* AGF#E-D, the full pentagram is only retrieved as such by TF-IDF (ranked 5th). The TF-IDF method also retrieves the 3 trigram and 2 tetragram substrings of this pattern, leading to redundancy in the output. The MGDP retrieves the 3 trigram substrings and SIA one. This recalls the fact that TF-IDF does not necessarily find minimal patterns, whereas MGDP is explicitly instructed not to consider longer patterns if a shorter one is already distinctive, and SIA finds

Table 4. *Centos* and retrieved patterns for *al-ḥiŷāz al-kabīr*, those of Chaachoo's centones that do not occur above the MFO are in parenthesis. Retrieved patterns in bold are exact matches to Chaachoo's, those underlined are substrings and those italicized are superstrings.

	Patterns for *al-ḥiŷāz al-kabīr*
Chaachoo's *centos*	AGF#E-D, F#GA, CBAG, (CED, FED)
TF-IDF	GF#E-, F#E-D, GF#E-D, AGF#E-, AGF#, **AGF#E-D**, GAGF#, *F#GAG*, **F#GA**, GAG, EF#G
SIA	AGF#, GAG, **F#GA**, BAG, CBA, EF#G, **CBAG**, DCB
MGDP	GF#E-, F#E-D, AGF#, DEF#, **F#GA**, EF#G

minimal patterns if they are not part of a longer pattern with the same frequency. However, SIA finds only one of the trigram substrings due to an artefact of the post-filtering process that has been applied (see Section 3.2), removing all non-contiguous MTP from the result set and removing patterns of length greater than 7. It turns out that the substrings GF#E- and F#E-D always appear, in this particular *ṭab'*, within a longer frequent pattern that is non-contiguous, and/or one that is longer than the maximum length limit.

The *centos* CED and FED (bracketed in Table 4) are missed by all methods as they occur below the minimum frequency. The SIA method alone finds the *cento* CBAG, split into the two overlapping trigrams CBA and BAG, but these are found by neither TF-IDF nor MGDP. For MGDP none of its substrings are distinctive above $\epsilon = 3$ (CBA: $\Delta = 2.10$; BAG: 2.78; CBAG: 2.31). These patterns can thus be found with a slight reduction of the distinctiveness threshold though with a larger set returned overall. The three patterns are discarded by TF-IDF, also due to low scores, all effectively zero.

Regarding some of the patterns which are not *centos* but are reported by one or more methods, SIA and TF-IDF report the pattern GAG which is frequent but not reported by MDGP, due to its Δ value of 1.74, below the minimum $\epsilon = 3.0$. The TF-IDF method also finds the extended GAGF# while MGDP and SIA find its suffix AGF#. As discussed above, all methods find the novel pattern EF#G. MGDP returns uniquely the pattern DEF#, and SIA uniquely the pattern DCB. These findings might correspond to recurrent *non-cento* patterns, such as orchestra performance preferences or expressive trends, and therefore open a path of further exploration.

6. Conclusions and future work

In this paper, we explore how computational methods can contribute to musicological research by focusing on the automatic analysis of melodic patterns in a music tradition that is under-represented in such studies, as is the case of Arab-Andalusian music. Drawing on theoretical formulations that are still in development by an expert performer and researcher of this tradition, Amin Chaachoo, we have tested three algorithms with the aim of learning how they perform in this specific music tradition and how the obtained results can contribute to the development of Chaachoo's theory. To implement this study, we have created a dataset of 145 machine-readable scores gathered from the Music Scores Collection of the CompMusic Arab-Andalusian Music Corpus, the largest source of machine readable data for the computational study of this music tradition.

From a musicological point of view, our research has helped deepen our understanding of the dataset, both in terms of its coverage and what the scores represent. Furthermore, through a discussion of the results with Chaachoo, we have been able to elicit theoretical information that was not completely developed in state of the art literature to inform the research presented here. Of special relevance amongst these outcomes is the identification of recurrent patterns that are not considered *centos*, such as orchestra preference and expressive trends. According to this finding, in future research, a search for recurrent patterns per orchestra should be performed, as well as on the whole dataset and per fundamental degree of the *ṭūbu'*, in order to find patterns related to expressive trends.

From a technical point of view, our methods are able to identify many of the frequent *centos* proposed by Chaachoo. It should be highlighted that in our research the list of these *centos* cannot be taken as a strict ground truth, but as a guide for studying the performance of the implemented technologies. This is in fact an interesting opportunity for developing strategies in which computational methods are used for exploratory analyses and discovery tasks such as the ones commonly addressed by musicologists, and is an element which we will need to emphasize in future work. Equally, from the discussions with Chaachoo, it seems that methods based on frequency are not sufficient for retrieving musically meaningful patterns and that some domain knowledge should be integrated in the task. Achieving this integration is an important target for future research.

Acknowledgments

The authors would like to thank Amin Chaachoo for his valuable work in creating and annotating the Arab-Andalusian Music Corpus and especially for his discussion of our results and the contributions to this paper that arose from that. The authors would also like to thank the CompMusic project for making the Arab-Andalusian Music Corpus publicly available, and Antoni Abelló Sanz for his work in manually segmenting the Music Scores Collection. Finally, the authors express gratitude to the reviewers for their well-considered and constructive feedback, crucial to the improving of this work.

Disclosure statement

No potential conflict of interest was reported by the author(s).

ORCID

Thomas Nuttall http://orcid.org/0000-0001-6316-1424
Andres Ferraro http://orcid.org/0000-0003-1236-2503
Darrell Conklin http://orcid.org/0000-0002-2313-9326
Rafael Caro Repetto http://orcid.org/0000-0003-2251-2202

References

Apel, W. 1958. *Gregorian Chant. A Midland Book*. Bloomington, Indianopolis: Indiana University Press.

Bozkurt, Barış, Ruhi Ayangil, and Andre Holzapfel. 2014. "Computational Analysis of Turkish Makam Music: Review of State-of-the-Art and Challenges." *Journal of New Music Research* 43 (1): 3–23. http://hdl.handle.net/10230/25935.

Caro Repetto, Rafael, Niccolo Pretto, Amin Chaachoo, Barış Bozkurt, and Xavier Serra. 2018. "An Open Corpus for the Computational Research of Arab-Andalusian Music." In *Proceedings of the 5th International Conference on Digital Libraries for Musicology (DLfM 2018)*, 78–86. Paris, France: Association for Computing Machinery. September 28. http://hdl.handle.net/10230/35470.

Chaachoo, A. 2011. *La música andalusí Al-Ála: Historia, conceptos y teoría musical*. Córdoba: Editorial Almuzara.

Chaachoo, A. 2016. *La musique hispano-arabe, al-Ala*. Univers musical. Editions L'Harmattan.

Chaachoo, A. 2019. *Al-qawā'id al-nazariyya lil-mūsīqā al-'andalusiyya al-maḡribiyya, al-'Āla (Theoretical principles of the Andalusian music from Morocco, al-Āla)*. Maṭbaa al-K̲alīj al-'Arabiyy.

Chewand, G., and J. W. McKinnon. 2001. "Centonization." *Oxford Music Online*.

Conklin, D. 2010. "Discovery of Distinctive Patterns in Music." *Intelligent Data Analysis* 14 (5): 547–554.

Ferraro, Andres, and Kjell Lemström. 2018. "On Large-Scale Genre Classification in Symbolically Encoded Music by Automatic Identification of Repeating Patterns." In *Proceedings of the 5th International Conference on Digital Libraries for Musicology DLfM '18*, 34–37. New York, NY, USA. Association for Computing Machinery. https://doi.org/10.1145/3273024.3273035.

Ferretti, P., and A. Agaësse. 1938. *Esthétique grégorienne ou Traité des formes musicales du chant grégorien. Volume I. Traduit de l'italien par Dom A. Agaësse*. Desclée.

Guettat, M. 2000. "La musique arabo-andalouse, l'empreinte du Maghreb." 560.

Janssen, B., W. De Haas, A. Volk, and P. Van Kranenburg. 2013. "Discovering Repeated Patterns in Music: State of Knowledge, Challenges, Perspectives." In *Proceedings of the 10th International Symposium on Computer Music Multidisciplinary Research*, Vol. 20. 74. Marseille, France.

Janssen, Berit, Peter van Kranenburg, and Anja Volk. 2015. "A Comparison of Symbolic Similarity Measures for Finding Occurrences of Melodic Segments." In *Proceedings of the 16th ISMIR Conference*, October 26–30, 659–665. Málaga, Spain: ISMIR press.

Meredith, David, Kjell Lemström, and Geraint A. Wiggins. 2002. "Algorithms for Discovering Repeated Patterns in Multidimensional Representations of Polyphonic Music." *Journal of New Music Research* 31 (4): 321–345. https://doi.org/10.1076/jnmr.31.4.321.14162.

Meredith, Dave, and Geraint A. Wiggins. 2001. "Pattern Induction and Matching in Polyphonic Music and Other Multi-Dimensional Datasets." In *In Callaos*, 61–66.

Nuttall, Thomas, Miguel García-Casado, Víctor Núñez-Tarifa, Rafael Caro Repetto, and Xavier Serra. 2019. "Contributing to New Musicological Theories with Computational Methods: The Case of Centonization in Arab-Andalusian Music." In *Proceedings of the 20th Conference of the International Society for Music Information Retrieval*, 04/11/2019, 223–228. Delft, The Netherlands. https://repositori.upf.edu/handle/10230/42789.

Pearson, Lara. 2016. "Coarticulation and Gesture: An Analysis of Melodic Movement in South Indian Raga Performance." *Music Analysis* 35 (3): 280–313. doi:10.1111/musa.12071.

Poché, C. 1997. *La música arábigo-andaluza (con CD)*. Músicas del mundo. Ediciones Akal.

Porter, A., M. Sordo, and X. Serra. 2013. "Dunya: A System for Browsing Audio Music Collections Exploiting Cultural Context." In *Proceedings of the 14th International Society for Music Information Retrieval Conference (ISMIR 2013)*, November 4, 101–106. Curitiba, Brazil: Audio Engineering Society. http://hdl.handle.net/10230/32251.

Ren, I. Y., H. V. Koops, A. Volk, and W. Swierstra. 2017. "In Search of the Consensus Among Musical Pattern Discovery Algorithms." In *Proceedings of the 18th International Society for Music Information Retrieval Conference, ISMIR 2017*, edited by Sally Jo Cunningham, Zhiyao Duan, Xiao Hu, and Douglas Turnbull, October 23–27, 671–678. Suzhou, China. https://ismir2017.smcnus.org/wp-content/uploads/2017/10/120_Paper.pdf.

Serra, Xavier. 2014. "Creating Research Corpora for the Computational Study of Music: the Case of the CompMusic Project." In *AES 53rd International Conference on Semantic Audio*, January 26, 1–9. AES. http://hdl.handle.net/10230/44221.

Sordo, Mohamed, Amin Chaachoo, and Xavier Serra. 2014. "Creating Corpora for Computational Research in Arab-Andalusian Music." In *Proceedings of the 1st International Workshop on Digital Libraries for Musicology DLfM '14*, 1–3. New York, NY, USA. Association for Computing Machinery. https://doi.org/10.1145/2660168.2660182.

Volk, Anja, and Peter van Kranenburg. 2001. "Melodic Similarity Among Folk Songs: An Annotation Study on Similarity-Based Categorization in Music." *Musicae Scientiae* 16 (3): 317–339.

Some observations on autocorrelated patterns within computational meter identification

Christopher Wm. White

The computational approach of autocorrelation relies on recurrent patterns within a musical signal to identify and analyze the meter of musical passages. This paper suggests that the autocorrelation process can act as a computational proxy for the act of period extraction, a crucial aspect of the cognition of musical meter, by identifying periodicities with which similar events tend to occur within a musical signal. Three analytical vignettes highlight three aspects of the identified patterns: (1) that the similarities between manifestations of the same patterns are often inexact, (2) that these patterns have ambiguous boundaries, and (3) that many more patterns exist on the musical surface than contribute to the passage's notated/felt meter, each of which overlaps with observations from music theory and behavioral research. An Online Supplement at chriswmwhite.com/autocorrelation contains accompanying data.

2010 Mathematics Subject Classifications: 00A65; 62P15; 91E10; 94A17
2012 Computing Classification Scheme: Applied computing Sound and music computing

1. Introduction

Meter is a phenomenon of patterns. In general, theorists imagine meter as arising from a series of consistently paced accents (Kozak 2020; Krebs 1999; Lerdahl and Jackendoff 1983; Repp 1998), as involving a listener who expects that pacing to continue into the future (Hasty 1997; London 2004), and as grouping adjacent pulses to form a hierarchy of stronger and weaker pulses (Cooper and Meyer 1960; Lerdahl and Jackendoff 1983), with quicker pulses evenly dividing broader pulses by two or three, creating consistent duple or triple relationships between levels (Cohn 2001, 2020). London (2004) suggests that the process of identifying a meter involves two complementary methods: (1) period extraction, and (2) template matching. The former – and more basic – strategy identifies patterns of recurrent periodicities present in some musical signal; the latter involves an approach that categorizes and organizes events into some a priori understanding of metric hierarchies and relationships.

Computational approaches to meter can broadly be seen to fall roughly along London's divide. The first camp includes those techniques that identify periodic patterns within some musical timeline, and then equates the patterns that emerge from those patterns with metrical pulses. This general approach includes such techniques as Discrete Fourier Transforms (Amiot 2016), wavelet analyses (Velarde, Meredith, and Weyde 2016), and resonant neural networks (Large,

Herrera, and Velasco 2015). However, perhaps the most straightforward and oldest of these approaches is *autocorrelation*, a technique that identifies the periods at which similar events appear in some musical timeline (Boone 2000; Brown 1993; Eck 2006; Eerola and Toiviainen 2004; Palmer and Krumhansl 1990; Volk 2008). While the basic method has been shown to identify the notated time signature of music in various styles and genres with a reasonably high level of precision (Gouyon et al. 2006; White 2019a), researchers have supplemented this technique with such additions as Expectation Maximization (de Haas and Volk 2016) and Shannon entropy (Eck and Casagrande 2005) to increase the process's utility and precision.

Much recent computational research can be seen as modeling the processes described by a template-matching approach (at least roughly). In many such approaches, some training session or encoded expert knowledge provides a model with expectations that allow it to categorize events into strong and weak beats, and to arrange those event into a metric hierarchy. Such approaches include context free grammars (McLeod and Steedman 2017; Rohrmeier 2020), hidden Markov models (Khadkevich et al. 2012; Papadopoulos and Peeters 2011), and deep learning techniques such as Support Vector Machines (Durand, David, and Richard 2014), recurrent neural networks (Böck, Krebs, and Widmer 2016; Durand et al. 2015), and temporal convolutional networks (Böck and Davies 2020). These approaches have proven to perform quite well against ground truth, and represent the current state of the art in meter detection within the music information retrieval literature.

However, while computational music research is often driven by the desire for an optimally efficacious model – that is, the best meter-finding model is the one whose output conforms to some expectation of desired results – such research may also be motivated by investigating the contours of the musical concept being modeled. It is with this latter purpose in mind that this paper approaches computational models of meter, and which motivates its study of a topic that, on the one hand constitutes the more elementary side of meter finding, while on the other hand represents a primary foundation of the theoretical concept of meter: period extraction. This study investigates some basic computational tactics underpinning period extraction, specifically analyzing the periodic patterns identified therein. While the specific mathematics and engineering behind the various approaches to computational period extraction substantially differ, the autocorrelation approach can be seen as representing the basic framework of many of these applications, as noted in Eck and Casagrande (2005) and Kim and Large (2017). This paper therefore adopts an autocorrelation approach to identify isochronous patterns in order to observe these patterns' attributes. Analyzing the musical characteristics of these patterns can then yield insights into the roles that period extraction and pattern finding potentially play within theoretical and cognitive concepts of meter. The broader goal of this study, then, is to provide an initial investigation into some intersections between computational, cognitive, and theoretical approaches to isochronous patterns within models of musical meter.

This paper will first offer a brief overview of the autocorrelation process, after which I will present three illustrative analytical vignettes using autocorrelation. I will highlight three particular aspects of the patterns identified in these analyses: (1) that the process often identifies patterns that are not particularly similar, but more similar than other options (or, *fuzzy patterns*), (2) that these patterns have ambiguous boundaries (or, the patterns are *unbounded*); and (3) that patterns exist on the musical surface in excess of those suggested by the notated/felt meter (or, *excessive patterns*). I then connect these aspects of computational periodic pattern identification to some broader theoretical and behavioral aspects of musical meter. While a thorough comparison with template-matching/categorization approaches is outside the boundaries of this paper, I end by situating this paper's results within the broader scope of computational approaches to meter finding.

For optimal transparency and accessibility, the computational implementations of each of the following analyses are presented as spreadsheets in this paper's Online Supplement. In this format, readers with no computational background can examine the engineering behind my analyses, and the procedures can be replicated or implemented in any computational environment with relative ease. These materials can be found at chriswmwhite.com/autocorrelation.

2. Autocorrelation: its engineering and its usage in meter finding

At its most basic, an autocorrelation approach to meter finding frames a musical passage as a timeline of values, with each consecutive point in that timeline representing an isochronous pulse (i.e. they are separated by the same time interval). Figure 1(a) shows a score for the opening measures of Fanny Mendelssohn Hensel's "Die Mainacht," while Figure 1(b) represents the events of each consecutive eighth-note pulse, and does so in three ways: as the number of note onsets at each moment, as the longest duration appearing at each moment (measured in eighth notes), and as the average pitch at each moment (measured in MIDI pitch, and counting only pitches that begin at each moment).

An autocorrelation method compares the initial – the prime – vector to each possible rotation of the timeline. Selected rotations appear under the note-onset timeline. Even visually, it is apparent that some of these rotations have more in common with the prime vector than do others. Rotating the vector by a single value – shifting each value by one-eighth note – mostly aligns zeros with non-zeros: this vector is very unlike the prime. However, because the note-onset patterns of both measures are identical, rotating the vector twelve eighth-notes (or, a full

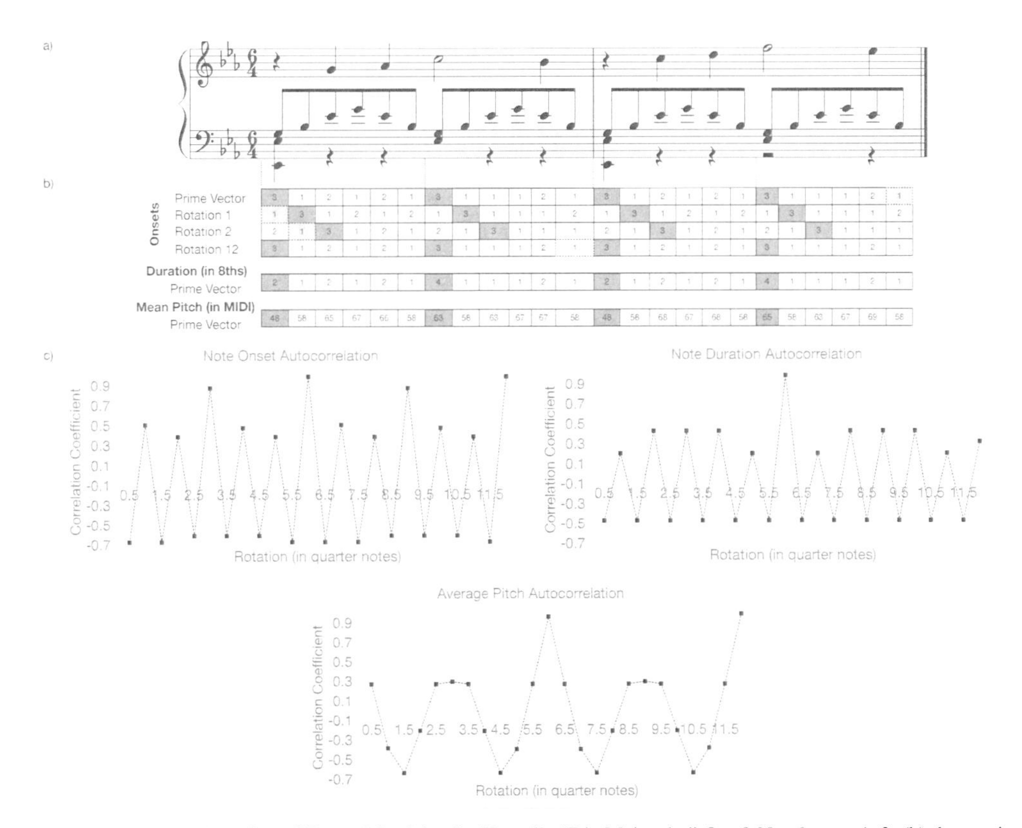

Figure 1. (a) An adaptation of Fanny Mendelssohn Hensel's "Die Mainacht," Op. 9 No. 6, mm. 1–2; (b) the music, represented as timelines of various parameters, such that each value corresponds to a consecutive eighth-note pulse, along with sample rotations; (c) the resulting autocorrelation values for each timeline.

measure) creates an identical correspondence. Autocorrelation quantifies these comparisons by correlating each rotation to its prime. For these calculations, a standard bivariate correlation is used (Pearson's r). Figure 1(c) shows the resulting correlation coefficients for all rotations for each parameter, with the x-axis metered such that each addition of 0.5 indicates a sequential rotation by eighth note. (NB: there can be some variation surrounding how to apply autocorrelation techniques; the approach outlined here rotates the whole length of the vector under consideration and consistently compares vectors of the same length to one another.)

To find the passage's meter, one can identify equally spaced peaks in the correlation coefficients, pinpointing isochronous patterns of consistent levels of similarity. In the current example, the most similarity occurs at the duration of the measure – events six quarter notes away from one another are quite similar. The note-onset and average-pitch timelines show the half-measure as the next most-similar isochronous periodicity, while the note-onset and duration timelines clearly show a quarter pulse. (The more-ambiguous quarter-note pulse of the average-pitch timeline is due to the stepwise motion of the melody and the skips to adjacent chord tones in the accompaniment: the average pitch heights of moments separated an eighth note are often quite similar in this texture!) In what follows, I use three musical excerpts to illustrate and describe the particular patterns identified by the procedure.

3. Three analytical vignettes: Some representative examples of meter finding using autocorrelation

3.1. *Procedure*

The following analyses use three parameters to construct timelines: note onsets, average pitch height, and duration. These three features are used because they have been shown to be salient to meter finding in both the cognitive and computational literature, with behavioral models showing that listeners associate a feeling of accent with relative peaks in note onsets (White 2019b), with certain durational patterns (Iversen, Patel, and Ohgushi 2008; Vos 1977; Woodrow 1951), as well as with changes in pitch contour (Acevedo, Temperley, and Pfordresher 2014; Huron and Royal 1996; Prince and Rice 2018; Thomassen 1982). Similarly, computational models often rely on durational patterns (McLeod and Steedman 2017; Nakamura et al. 2017) and note-onset data (Gouyon and Dixon 2005; Gouyon et al. 2006; this parameter is generally used when analyzing symbolic score data, as onsets can serve as a proxy for loudness). Additionally, spectrogram models implicitly involve information about pitch distributions (e.g. Böck and Davies 2020; admittedly, the pitch information involved in a spectrogram will be more nuanced than the symbolic pitch data used here). Again, note that my goal here is simply to illustrate and study the patterns found within a period-extraction model: other musical features like harmonic change, melodic leaps, or phrasing – or the combination of multiple features – could also participate in the meter of these passages.

The process as implemented here identifies the shortest note duration, divides the pieces into consecutive events of that duration, and creates a timeline such that each value corresponds to each consecutive slice (this duration was the eighth note in Figure 1, for instance). For simplicity, if the quickest duration of the passage only occurs a handful of times (as will be the case for the sixteenth-note pulse in Figure 2) or has an ambiguous duration (as will be the case for the grace notes in Figure 3), the next-quickest pulse is used. The three approaches are implemented as follows: the note onset approach shows the number of notes that initiate/begin at each pulse, with zero indicating either rests or sustained events. The average-pitch approach associates MIDI numbers with each pitch onset, and averages these numbers; sustained notes do not contribute to the average pitch height. The duration approach represents the length of the longest note

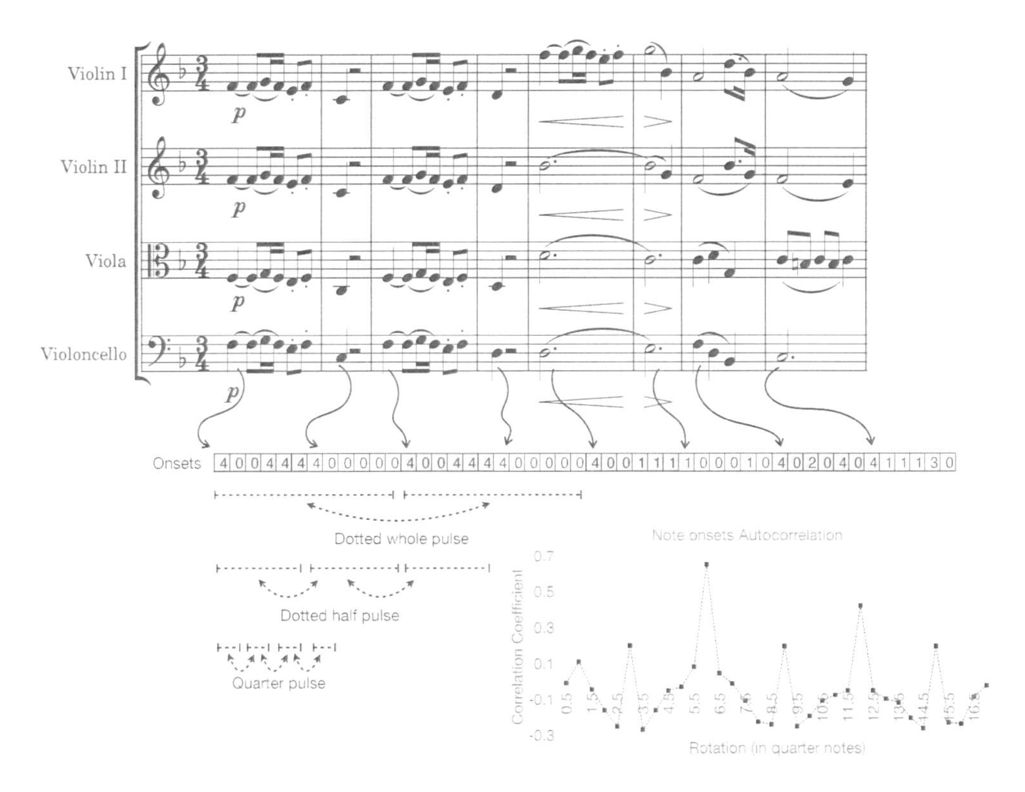

Figure 2. Ludwig v. Beethoven, String Quartet, Op 18, no 1, mm. 1–8 and accompanying autocorrelations.

begun at each timepoint (with the duration generating the timeline assigned a value of 1). For simplicity and brevity, the following analyses presents a selection of these approaches, tailored to the characteristics of each example. Finally, given that this paper is interrogating the patterns used in these processes rather than advocating for a particular approach, I do not implement a rigorous meter-finding or time-signature-identification step but rather simply comment on the periodicities that would be available to such a process.

3.2. *Vignette 1: Fuzzy patterns in Ludwig v. Beethoven's String Quartet, op. 18, no 1*

Figure 2 shows the opening to Ludwig v. Beethoven's String Quartet opus 18, number 1, with its corresponding note-onset vector. Autocorrelation identifies a triple grouping of quarter notes, and duple groupings of 16th notes and 8th notes: these are precisely the periodicities indicated by the time signature. It also identifies a sextuple and 24-tet grouping of quarter notes, thus identifying a two-measure hypermeter, and the eight-measure phrase. These groupings are noted by the brackets underneath the timeline.

Several of these identified patterns are straightforward and visually apparent. The fact that several motives explicitly return (mm. 1 and 3, mm. 9 and 11) and several rhythmic profiles return (mm. 2 and 4, mm. 6 and 8, 11 and 13, mm. 14 and 16, mm. 15 and 17) results in the a two-bar hypermetric patterns. The duple divisions of the quarter-note pulse are also straightforward to identify: more events occur on the quarter pulses than on the alternating eighth pulses. The dotted-half-note patterns, however, are less immediately apparent, but are still present. The music's recurrent motives tend to place events with longer durations at the beginnings of

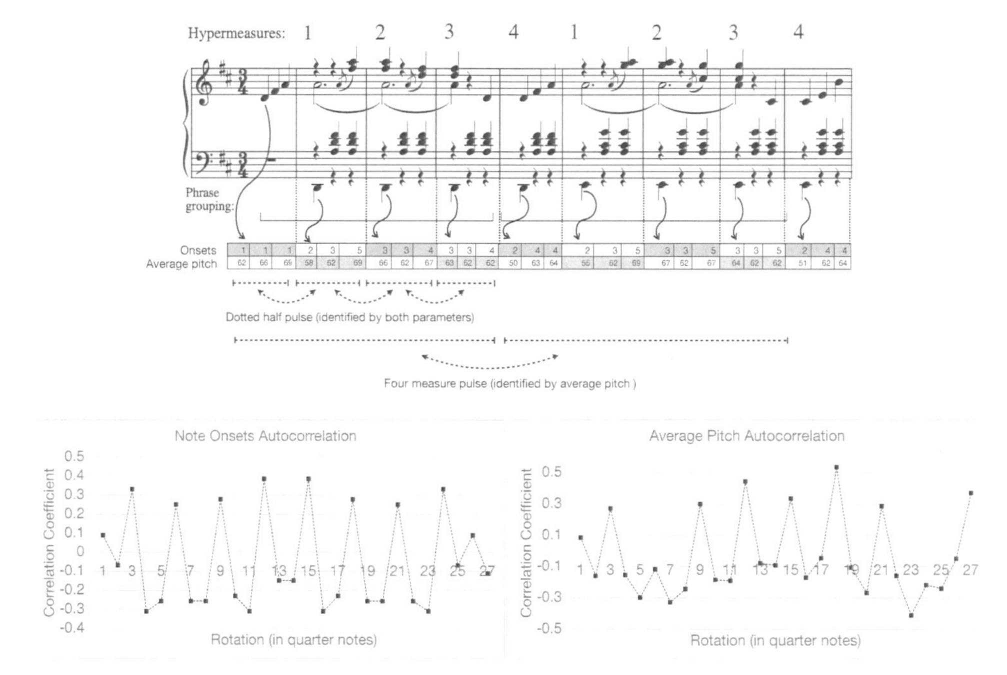

Figure 3. J. Strauss II, An der schnen blauen Donau, Op. 314, mm. 19, with annotations from the analysis of Rothstein 1989, 4–7.

measures while events with shorter durations appear at the ends of measures, thereby creating patterns on onsets whose density fluctuations at the dotted-quarter pulse.

This dotted-quarter pulse exhibits what I call *fuzzy* patterns (roughly adapting the adjective's usage in Quinn, 2001). The autocorrelation procedure illustrates peaks at that pulse not because all – or even most – events at the remove of that duration are similar, but because more similarities arise at that pulse duration than at other durations (while the rotations of 3 and 9 and 15 quarter notes return a coefficient of only 0.2, the surrounding values are considerably lower).

3.3. *Vignette 2: Unbounded patterns in J. Straus's "An der schnen blauen Donau"*

Figure 3 illustrates a piano reduction of the opening to the Johann Straus waltz "An der schnen blauen Donau", with timelines indicating note onsets and average pitch height, now divided into each quarter pulse. The patterns of note onsets produced by the "Oom-pa-pa" oscillations of the waltz provide a clear triple pattern; the relatively low pitches that initiate each measure create a similar three-quarter pattern within the average pitch height timeline. The melody's consistent shift between a treble register (mm. 1, 5, 9) and a soprano register (mm. 2–4, 6–8) provides a further four-measure pattern within that approach.

This example provides further instances of fuzzy patterns returned within this passage: a rotation of two-measures aligns events like the rising three-quarter-note gestures of measures 1 and 5 with the characteristic Viennese accents on the first and third beats of measures 3 and 7, resulting in a low correlation (a $-.17$ correlation coefficient, shown in the supplement). Again, however, this alignment produces a correlation better than the surrounding quarter notes. Such local maxima within otherwise-low coefficients can combine with spikes at the third and ninth rotations to express the dotted-quarter pulse.

More notable, however, is that the arrangement of correlations does not indicate that the first measure would be felt as an upbeat to the proceeding measures. Note that Figure 3 includes

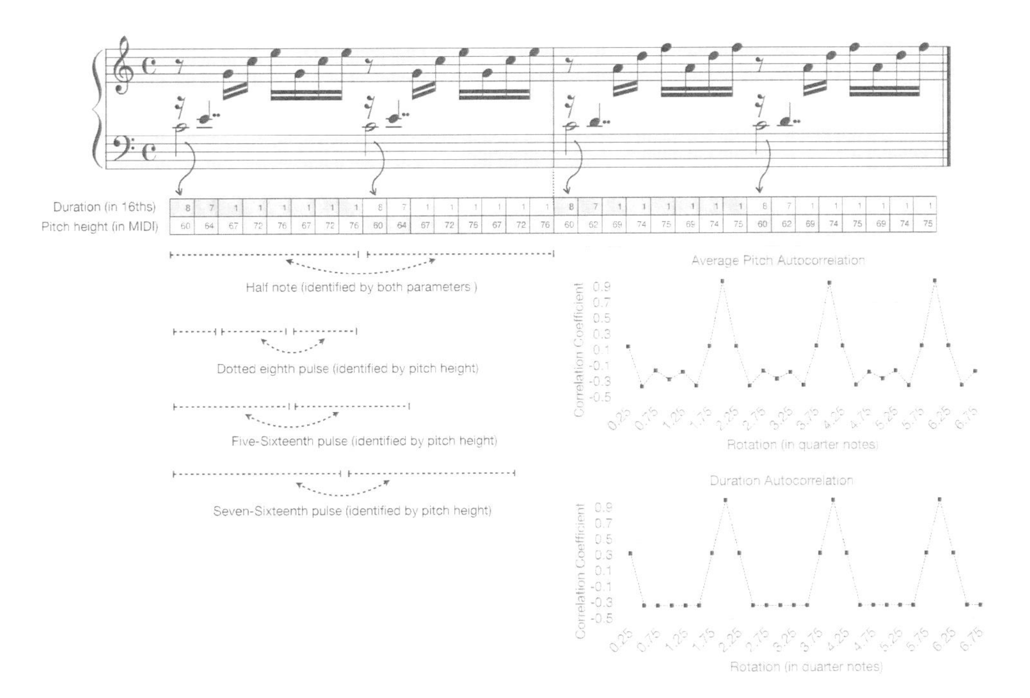

Figure 4. J.S. Bach's Well Tempered Clavier, Book 1, Prelude 1, mm. 1–2.

hypermetric and phrasing annotations for the passage (drawn from Rothstein 1989) which show that – while there exists a four-bar hypermeter and a four-bar phrase – the phrase and hypermeter are out of phase, such that the first measure is felt as a hyper-upbeat to the second. The autocorrelations are agnostic to this difference, indicating only that there is a four-bar hyper-measure, not where it begins. These patterns simply rely on their periodic similarities rather than specific boundaries that denote their beginnings and endings – these patterns are unbounded. Below, I'll return to the implications of unbounded patterns as regards this procedure's identification of meter. (The "unboundedness" of the patterns in somewhat obscured in the orthography of the figures, as the brackets have beginnings and endings. It is useful to recall that, for instance, two adjacent measure-long brackets simply illustrate that events removed at the distance of a measure in that passage are similar.)

3.4. *Vignette 3: Excessive patterns in Johann S. Bach's WTC 1, no. 1*

Figure 4 shows the first two measures of Prelude 1 from Johann S. Bach's *Well Tempered Clavier*, Book 1, along with duration and pitch timelines with sixteenth-note divisions. Both timelines identify periodicities at the half-note pulse, a predictable result given that Bach's figuration pattern repeats every half measure. What is more notable, however, are the peaks in correlation at 3, 5, and 7 rotations of the pitch-height timeline, indicating similar patterns of events at the remove of the dotted-eighth, five-sixteenth, and seventh-sixteenth periodicities. Similar peaks occur at the same relative points corresponding to each measure within the correlation graphs.

This unusual series of peaks is due to the repeated figuration arpeggiating the underlying chord: in each half-measure pattern, the initial two sixteenth notes are followed by two repetitions of the same three-note pitch pattern. Because of this, a rotation of three sixteenths will exactly align this trio of pitches with one another; similarly, a rotation of five sixteenth will align the second triple group in the first figure (the 6th, 7th, and 8th events in the prime timeline) with the first triple group in the second figure (the 11th, 12th, and 13th events). Additionally, exemplifying a more fuzzy pattern, the first five and last three notes of the figure produce consistently rising contours, and rotating the vector at the remove of those corresponding patterns produces higher correlations than do rotations of the surrounding values. (Notably, this analysis resonates with the 2+3+3 non-isochronous accent pattern with which Cone (1968) reads in this piece.)

These patterns do not align with the notated – and likely perceived – prevailing duple meter of this passage. A human analyst would would almost certainly discard these triple, quintuple, and septuple patterns as non-metrical, either because their own divisions would be non-isochronous (i.e. they mix duple and triple groups), because they do not evenly divide the clear half and whole-note pulses, or because they are insufficiently regular (they "re-begin" in each measure). However, these patterns do exist in this music, and a local, pitch-based understanding of this passage would notice such patterns. These patterns appear to exist in excess of metrical frameworks. Such *excessive* patterns are not uncommon within autocorrelated musical timelines (White 2019a), and suggest that a musical surface includes more isochronous patterns than contribute to the prevailing meter.

4. Pattern finding and musical meter

The patterns outlined offer overlaps with and insights into theoretical and cognitive conceptions of meter. In what follows, I outline some of these intersections, particularly in how fuzzy, unbounded, and excessive patterns function in these analyses.

4.1. *Fuzzy patterns*

These examples consistently feature fuzzy patterns, or patterns whose periodic similarities are identified simply because they were more similar than other periodicities. On the one hand, such patterns would seem to align with what we know about humans' capacity to project metric periodicities onto musical stimuli, even if the stimuli provide very sparse cues (Dawe, Platt, and Racine 1994; Toiviainen and Snyder 2003), conflicting/competing cues (Pfordresher 2003; Prince, Thompson, and Schmuckler 2009; Prince and Rice 2018; Repp 2007), and even no cues (Bolton 1894; Brochard et al. 2003). Furthermore, once a listener has entrained to a metric pattern, they have the capacity to continue hearing a metric pattern even in the face of conflicting or sparse information (Krebs 1999; Lerdahl and Jackendoff 1983; London 2004).

On the other hand, fuzzy patterns also have the potential to be at odds with perceived meter. There are limits to the listeners' capacities to identify and entrain to patterns in music, be the patterns too subtle to notice (Fraisse 1946; Szelag et al. 1998; White 2017; Woodrow 1909) or in conflict with a pattern from another musical domain (Ellis and Jones 2009; London, Himberg, and Cross 2009; Prince, Thompson, and Schmuckler 2009). The subtlety of some of the patterns in the Figure 4, for instance, would seem difficult – if not impossible – to hear. While a computational approach, then, might focus on identifying any pattern present within a musical timeline, some such patterns might be *too fuzzy* for a cognitive model. To align a pattern-finding model of musical meter with cognition, then, one would need to import some definition of what is "too fuzzy" to be perceived by a human listener to contribute to a felt meter.

4.2. *Unbounded patterns*

As observed in Eck and Casagrande (2005), any model based solely on recurrent patterns will not indicate the location of relatively strong beats, or even where a metric pattern starts. The pattern-finding process simply identifies that events removed by some consistent periodicity are similar. It does not say where those patterns begin. Recall that Figure 3 included two parallel annotations, that of the four-bar phrase, and that of the four-bar hypermeter, the former beginning in measure 1, the latter beginning in measure 2. A pattern-finding procedure like autocorrelation does not distinguish between these two readings: it simply indicates that events are quite similar to events that occur at the distance of four measures. In order to distinguish the two parsings of this passage, the process would need some definition of "accent" or "beginning," and such definitions are not available to a model that engages purely in period extraction. In other words, any meter-identifying process that focuses solely on isochronous patterns might be able to identify the periods with which events recur in some signal, but it will not be able to assign a downbeat within that period without some further definition of what constitutes a downbeat. I return to this issue in my final comparison with template-matching procedures. It should be noted that Eck and Casagrande (2005) do indeed devise an adaptation to autocorrelation that identifies the initiation of patterns; however, their additions result in their overall method being a hybrid between period extraction and template matching. Their adaptation uses Shannon entropy to equate more contextually unusual/less predictable events with metrical stress, something which can be seen to import a definition of accent to categorize strong and weak pulse layers.

4.3. *Excessive patterns*

In Figure 4, the triple groupings embedded within the repeated 8-note figure produced several peaks at periodicities that conflict with the notated duple meter; as noted above, an analyst or listener would find no shortage of reasons to discard these patterns in favor of a purely duple reading. But such patterns are present on the musical surface, and they do pervade the passage. As has been demonstrated elsewhere (Conklin 2010; Jehan 2005; White 2019a), discovering patterns on a musical surface poses little problem for a computational model of meter; rather, choosing which patterns to use in the final assessment presents a challenge. The excessive patterns returned by period extraction also demonstrates a notable interconnection between meter and other types of musical patterns by highlighting the porous boundary between metric patterns and motivic/thematic groups. Indeed, while this paper highlighted the role of automated pattern finding for meter identification, the identification of recurrent strings of similar musical events has aided the study of piece-specific and style-generic event sequences (Cambouropoulos 2001; Conklin 2010) as well as musical form (Bañuelos and Orduña 2017). Indeed, researchers have studied the grey area that exists between recurrent patterns that participate in a meter and those that are simply thematic, being the focus of both music theorists (Krebs 1999; Riemann 1903) and cognitive researchers (Acevedo, Temperley, and Pfordresher 2014; Prince 2014). The output of a period-extracting model like autocorrelation explicitly shows the shared resources used by (and definitional overlap between) both metric and motivic musical patterns.

4.4. *Periodic pattern finding and meter*

In my initial discussion, I invoked London's dichotomy between period extraction and template matching as a way to heuristically divide computational approaches into those that rely on identifying surface patterns versus those that categorize pulses and events. I also noted that isochronous pattern-finding models have generally been shown to underperform models that involve some manner of expert knowledge or training session. I end this paper by considering what this apparent imbalance might tell us about the nature of meter and the cognitive act of meter finding. First, as observed in Jehan (2005), "downbeat estimation requires some fair amount of prior knowledge," and the simple act of period extraction lacks this prior knowledge. This study demonstrates exactly why this prior knowledge is required, as models such as those used in the current paper lack the ability to (a) provide boundaries on the kinds of imprecise and fuzzy patterns that are considered salient, (b) include some definition of accent or initiation to distinguish between levels of metric strength, and (c) select from the array of possible patterns in such a way that ensures sufficient regularity and the duple and triple relationships between pulses. Periodicities on the musical surface are therefore necessary but not sufficient when determining the perceived or intended meter of a musical passage. While a full review and comparison of template-matching/categorization models is outside the purview of this paper, this insufficiency can be rectified by pre-programming a model with some definition of accent (McLeod and Steedman 2017) or including a deep-learning training session (Böck and Davies 2020).

And so: what of period extraction and pattern finding? Regardless of its relative efficacy, isolating periodic patterns within a meter finding task demonstrates the role these patterns play in the larger concept of meter. At the broadest level, this study suggests that meter is a phenomenon of interpretation and organization: while isochronous and recurrent patterns may pervade some musical signal, they remain insufficient to model musical meter on their own. More specifically: modeling period extraction allows for a closer examination of characteristics of recurrent patterns that contribute to musical meter, yielding – for instance – insights into how those patterns overlap with notions of motivic grouping. More speculatively: this study's suggestion that period extraction is insufficient to identify a piece's meter in and of itself has implications for cognitive and embodied theories of metric entrainment. And yet, a cognitive model of meter *is* a model reliant on isochronous and periodic musical patterns – from a listener's standpoint, meter allows for the organization of events into equally spaced patterns that both explain past events and predict future ones (Hasty 1997; Kozak 2020; London 2004). If a listener's experience of a musical meter involves *both* period extraction *and* template matching, this study suggests that the former is crucially reliant on the latter to produce a coherent metric assessment.

Acknowledgments

I thank Darrell Conklin for his guidance and feedback during this project's development. I would also like to thank the two anonymous reviewers for their invaluable feedback on earlier drafts.

Disclosure statement

No potential conflict of interest was reported by the author(s).

Supplemental online material

Supplemental data for this article can be accessed online at https://doi.org/10.1080/17459737.2021.1923843.

ORCID

Christopher Wm. White http://orcid.org/0000-0002-6435-6423

References

Acevedo, Stefanie, David Temperley, and Peter Q. Pfordresher. 2014. "Effects of Metrical Encoding on Melody Recognition." *Music Perception: An Interdisciplinary Journal* 31 (4): 372–386.

Amiot, Emmanuel. 2016. *Music through Fourier Space: Discrete Fourier Transform in Music Theory*. Heidelberg: Springer.

Bañuelos, Cristian, and Felipe Orduña. 2017. "Dynamic Time Warping for Automatic Musical Form Identification in Symbolical Music Files." In *Mathematics and Computation in Music*, edited by Octavio A. Agustín-Aquino, Emilio Lluis-Puebla, and Mariana Montiel, 253–258. Cham: Springer International Publishing.
Böck, Sebastian, and Matthew Davies. 2020. "Deconstruct, Analyse, Reconstruct: How to Improve Tempo, Beat, and Downbeat Estimation." In *Proceedings of the 21st International Society for Music Information Retrieval Conference (virtual)*, 574–582.
Böck, S., F. Krebs, and G. Widmer. 2016. "Joint Beat and Downbeat Tracking with Recurrent Neural Networks." In *Proceedings of the 17th International Society for Music Information Retrieval*. New York.
Bolton, T. L. 1894. "Rhythm." *The American Journal of Psychology* 6 (2): 145–238.
Boone, G. M. 2000. "Marking Mensural Time." *Music Theory Spectrum* 22 (1): 1–43.
Brochard, R., D. Abecasis, D. Potter, R. Ragot, and C. Drake. 2003. "The 'Ticktock' of Our Internal Clock: Direct Brain Evidence of Subjective Accents in Isochronous Sequences." *Psychological Science* 14, 362–366.
Brown, J. C. 1993. "The Determination of Meter of Musical Scores by Autocorrelation." *Journal of the Acoustic Society of America* 94 (4): 1953–1957.
Cambouropoulos, Emilios. 2001. "Melodic Cue Abstraction, Similarity, and Category Formation: A Formal Model." *Music Perception* 18 (3): 347–370.
Cohn, Richard. 2001. "Complex Hemiolas, Ski-hill Graphs and Metric Spaces." *Music Analysis* 20 (3): 295–326.
Cohn, Richard. 2020. "Meter." In *The Oxford Handbook of Critical Concepts in Music Theory*, edited by Alex Rehding and Steven Rings, 207–233. New York: Oxford University Press.
Cone, Edward T. 1968. *Music through Fourier Space: Discrete Fourier Transform in Music Theory*. New York: W.W. Norton.
Conklin, Darrell. 2010. "Discovery of Distinctive Patterns in Music." *Intelligent Data Analysis* 14, 547–554.
Cooper, Grosvenor, and Leonard Meyer. 1960. *The Rhythmic Structure of Music*. Chicago, IL: The University of Chicago Press.
Dawe, L. A., J. R. Platt, and R. J. Racine. 1994. "Inference of Metrical Structure from Perception of Iterative Pulses Within Time Spans Defined by Chord Changes." *Music Perception* 12 (1): 57–76.
de Haas, W. B., and A. Volk. 2016. "Meter Detection in Symbolic Music Using Inner Metric Analysis." In *Proceedings of the 17th Conference of the International Society for Music Information Retrieval, New York, USA*, 574–582.
Durand, S., J. P. Bello, B. David, and G. Richard. 2015. "Downbeat Tracking with Multiple Features and Deep Neural Networks." In *2015 IEEE International Conference on Acoustics, Speech and Signal Processing (ICASSP)*, Brisbane, 409–413.
Durand, S., B. David, and G. Richard. 2014. "Enhancing Downbeat Detection When Facing Different Music Styles." In *2014 IEEE International Conference on Acoustics, Speech and Signal Processing (ICASSP)*, Florence, 3132–3136.
Eck, D. 2006. "Identifying Metrical and Temporal Structure with An Autocorrelation Phase Matrix." *Music Perception* 24 (2): 167–176.
Eck, Douglas, and Norman Casagrande. 2005. "Finding Meter in Music Using An Autocorrelation Phase Matrix and Shannon Entropy." In *ISMIR 2005 – 6th International Conference on Music Information Retrieval*, London, 01, 504–509.
Eerola, Tuomas, and Petri Toiviainen. 2004. "MIR in Matlab: The MIDI Toolbox." In *Proceeding of 5th International Conference on Music Information Retrieval, Barcelona*.
Ellis, R. J., and M. R. Jones. 2009. "The Role of Accent Salience and Joint Accent Structure in Meter Perception." *Journal of Experimental Psychology: Human Perception and Performance* 35, 264–280.
Fraisse, P. 1946. "Contribution à l'étude du rhythme en tant que forme temporelle." *Journal de Psychologie Normale et Pathologique* 39, 283–304.
Gouyon, Fabien, and Simon Dixon. 2005. "A Review of Automatic Rhythm Description Systems." *Computer Music Journal* 29, 34–54.
Gouyon, Fabien, Gerhard Widmer, Xavier Serra, and Arthur Flexer. 2006. "Acoustic Cues to Beat Induction: A Machine Learning Perspective." *Music Perception: An Interdisciplinary Journal* 24 (2): 177–188.
Hasty, Christopher F. 1997. *Meter as Rhythm*. New York: Oxford University Press.
Huron, David, and Matthew Royal. 1996. "What Is Melodic Accent? Converging Evidence From Musical Practice." *Music Perception: An Interdisciplinary Journal* 13 (4): 489–516.

Iversen, John R., Aniruddh D. Patel, and Kengo Ohgushi. 2008. "Perception of Rhythmic Grouping Depends on Auditory Experience." *The Journal of the Acoustical Society of America* 124 (4): 2263–2271. doi:10.1121/1.2973189

Jehan, Tristan. 2005. "Downbeat Prediction by Listening and Learning." In *IEEE Workshop on Applications of Signal Processing to Audio and Acoustics*, New Paltz, 11, 267–270.

Khadkevich, M., T. Fillon, G. Richard, and M. Omologo. 2012. "A Probabilistic Approach to Simultaneous Extraction of Beats and Downbeats." In *2012 IEEE International Conference on Acoustics, Speech and Signal Processing (ICASSP)*, Kyoto, 445–448.

Kim, J. C., and Edward Large. 2017. "Entrainment of Canonical Oscillators to Complex Rhythms: Temporal Receptive Field Revisited." Paper presented at the Society for Music Perception and Cognition meeting, San Diego.

Kozak, Mariusz. 2020. *Enacting Musical Time: The Bodily Experience of New Music*. New York: Oxford University Press.

Krebs, Harold. 1999. *Fantasy Pieces: Metrical Dissonance in the Music of Robert Schumann*. New York: Oxford University Press.

Large, E. W., J. A. Herrera, and M. J. Velasco. 2015. "Neural Networks for Beat Perception in Musical Rhythm." *Frontiers in Systems Neuroscience* 9, 159–173.

Lerdahl, Fred, and Ray Jackendoff. 1983. *A Generative Theory of Tonal Music*. Cambridge, MA: MIT Press.

London, Justin. 2004. *Hearing in Time: Psychological Aspects of Musical Meter*. New York: Oxford University Press.

London, Justin, Tommi Himberg, and Ian Cross. 2009. "The Effect of Structural and Performance Factors in the Perception of Anacruses." *Music Perception: An Interdisciplinary Journal* 27 (2): 103–120.

McLeod, Andrew, and M. Steedman. 2017. "Meter Detection in Symbolic Music Using a Lexicalized PCFG." In *Proceedings of the 14th Sound and Music Computing Conference*, Espoo, Finland, 373–379.

Nakamura, Eita, Kazuyoshi Yoshii, Shigeki Sagayama, Eita Nakamura, Kazuyoshi Yoshii, and Shigeki Sagayama. 2017. "Rhythm Transcription of Polyphonic Piano Music Based on Merged-Output HMM for Multiple Voices." *IEEE/ACM Transactions on Audio, Speech, and Language Processing* 25 (4): 794–806. doi:10.1109/TASLP.2017.2662479

Palmer, C., and C. L. Krumhansl. 1990. "Mental Representations for Musical Meter." *Journal of Experimental Psychology: Human Perception and Performance* 16, 728–741.

Papadopoulos, H., and G. Peeters. 2011. "Joint Estimation of Chords and Downbeats From An Audio Signal." *IEEE Transactions on Audio, Speech, and Language Processing* 19 (1): 138–152.

Pfordresher, Peter. 2003. "The Role of Melodic and Rhythmic Accents in Musical Structure." *Music Perception* 20, 431–464.

Prince, Jon. 2014. "Pitch Structure, But Not Selective Attention, Affects Accent Weightings in Metrical Grouping." *Journal of Experimental Psychology: Human Perception and Performance* 40, 2073–2090.

Prince, Jon B., and Tim Rice. 2018. "Effects of Metrical Encoding on Melody Recognition." *Journal of Experimental Psychology: Human Perception and Performance* 44, 1356–1367.

Prince, Jon B., William F. Thompson, and Mark A. Schmuckler. 2009. "Pitch and Time, Tonality and Meter: How Do Musical Dimensions Combine?" *Journal of Experimental Psychology* 35 (5): 1598–1617.

Repp, Bruno H. 1998. "Variations on a Theme by Chopin: Relations Between Perception and Production of Timing in Music." *Journal of Experimental Psychology: Human Perception and Performance* 24, 791–811.

Repp, Bruno H. 2007. "Hearing a Melody in Different Ways: Multistability of Metrical Interpretation, Reflected in Rate Limits of Sensorimotor Synchronization." *Cognition* 102 (3): 434–454.

Riemann, Hugo. 1903. *System der Musikalischen Rhythmik und Metrik*. Leipzig: Breitkopf und Haertel.

Rohrmeier, Martin. 2020. "Towards a Formalization of Musical Rhythm." In *Proceedings of the 21st International Society for Music Information Retrieval Conference (virtual)*, 621–629.

Rothstein, William N. 1989. *Phrase Rhythm in Tonal Music*. New York: Schirmer.

Szelag, E., J. Kowalska, K. Rymarczyk, and E. Pöppel. 1998. "Temporal Integration in a Subjective Accentuation Task As a Function of Child Cognitive Development." *Neuroscience Letters* 257 (2): 69–72.

Thomassen, Joseph M. 1982. "Melodic Accent: Experiments and a Tentative Model." *The Journal of the Acoustical Society of America* 71 (6): 1596–1605. doi:10.1121/1.387814

Toiviainen, P., and J. S. Snyder. 2003. "Tapping to Bach: Resonance-Based Modeling of Pulse." *Music Perception* 21 (1): 43–80.

Velarde, Gissel, David Meredith, and Tillman Weyde. 2016. *A Wavelet-Based Approach to Pattern Discovery in Melodies*. In *Computational Music Analysis*, edited by D. Meredith, 303–333. Cham: Springer.
Volk, Anja. 2008. "The Study of Syncopation Using Inner Metric Analysis: Linking Theoretical and Experimental Analysis of Metre in Music." *Journal of New Music Research* 37 (4): 259–273.
Vos, P. 1977. "Temporal Duration Factors in the Perception of Auditory Rhythmic Patterns." *IScientific Aesthetics/Sciences de l'Art* 1, 183–199.
White, Christopher Wm. 2017. "Relationships Between Tonal Stability and Metrical Accent in Monophonic Contexts." *Empirical Musicology Review* 12 (1): 19–27.
White, Christopher Wm. 2019a. "Autocorrelation of Pitch-Event Vectors in Meter Finding." In *Mathematics and Computation in Music, Proceedings of the 7th International Conference*, edited by Mariana Montiel, Francisco Gómez-Martín, and Octavio A. Agustín-Aquino. Lecture Notes in Computer Science, 287–296. Heidelberg: Springer.
White, Christopher Wm. 2019b. "Influences of Chord Change on Metric Accent." *Psychomusicology: Music, Mind, and Brain* 29 (4): 209–225.
Woodrow, H. 1909. "A Quantitative Study of Rhythm." *Archives of Psychology* 18, 1–66.
Woodrow, H. 1951. "Time Perception." In *Handbook of Experimental Psychology*, edited by S. S. Stevens, 1224–1236. New York, NY: Wiley.

Exploring annotations for musical pattern discovery gathered with digital annotation tools

Darian Tomašević, Stephan Wells, Iris Yuping Ren, Anja Volk, and Matevž Pesek

The study of inter-annotator agreement in musical pattern annotations has gained increased attention over the past few years. While expert annotations are often taken as the reference for evaluating pattern discovery algorithms, relying on just one reference is not usually sufficient to capture the complex musical relations between patterns. In this paper, we address the potential of digital annotation tools to enable large-scale annotations of musical patterns, by comparing datasets gathered with two recently developed digital tools. We investigate the influence of the tools and different annotator backgrounds on the annotation process by performing inter-annotator agreement analysis and feature-based analysis on the annotated patterns. We discuss implications for further adaptation of annotation tools, and the potential for deriving reference data from such rich annotation datasets for the evaluation of automatic pattern discovery algorithms in the future.

2010 Mathematics Subject Classifications: 00A65; 97R50

1. Introduction

A significant number of topics currently researched in the field of Music Information Retrieval (MIR) rely heavily on using reference or "ground truth" data, often derived from human annotations, for the evaluation of algorithms. The comparison of state-of-the-art algorithms on the different tasks in the yearly rounds of the MIREX (The Music Information Retrieval Evaluation eXchange) framework has uncovered the issue of ambiguity of musical structures for evaluating algorithms, most notably by uncovering differences in annotations. For instance, Flexer and Grill (2016) discovered a rather low inter-annotator agreement for the MIREX music similarity task, unveiling the problem of using a single-reference annotation for evaluating similarity algorithms. Koops et al. (2019) reached similar conclusions for the chord estimation task, showing low inter-annotator agreement for chord annotations between musical experts. Furthermore, Balke et al. (2016) showed how the evaluation of automatic melody finding algorithms depends heavily on the choice of the human annotator for providing the ground truth.

The automatic discovery of musical patterns has been a long standing research topic in computational music analysis (Janssen et al. 2013), evolving into the MIREX task termed Discovery of Repeated Themes & Sections (Collins 2011, 2019; Ren, Volk, et al. 2018). In this task, the evaluation of newly-proposed algorithms is carried out with reference annotations based on music-theoretic analyses by three experts. However, the ambiguity of musical structures and the different conceptualisations of the notion of patterns make the use of reference data relying on only a very small number of experts rather problematic: there is no clear single comprehensive definition of what constitutes a pattern, or even repetition (Taube 1995; Margulis 2014; Collins 2019; Melkonian et al. 2019; Sears and Widmer 2020). For example, a musical pattern can be described as "a salient recurring figure or short musical idea of special importance" (Nieto and Farbood 2012), while the MIREX task defines patterns as a "set of ontime-pitch pairs that occur at least twice (i.e. is repeated at least once) in a piece of music" (Collins 2019). Moreover, not all recurring sequences are perceived as patterns by the listener, as this depends on the structural position of the pattern (Margulis 2014), the listener's moment-to-moment perception, and other influencing factors such as the listener's musical background or music theoretic education. Automatically discovered patterns do not have to be perceivable, if they are useful in other contexts, such as for supporting automatic composition (Herremans and Chew 2017) or classification (Boot, Volk, and de Haas 2016). However, in other contexts it is important that automatically detected patterns are perceivable for listeners, such as in music education (Bamberger 2000).

Gathering annotations from different listeners on the same pieces therefore allows the study of differences and commonalities between listeners regarding their conceptualisation of patterns. Such a study can pave the way for a more valid evaluation of algorithms, based on the consideration of commonalities and differences between listeners or groups of listeners, instead of a very small number of annotators. The issue of inter-annotator agreement in pattern discovery was previously addressed by Nieto and Farbood (2012) through gathering multiple annotations for a single dataset from fourteen annotators on six music excerpts using pen and paper. This dataset was later digitised as the HEMAN (Human Estimations of Musically Agreeing Notes) dataset in Ren, Koops, et al. (2018). In analysing the patterns, considerable disagreement between the annotators was discovered. This disagreement could be reduced by considering the relevance/confidence scores of patterns marked by the annotators and by lowering the time resolution to allow for more tolerance towards differences in start and end times of annotations.

While pen-and-paper annotation has been most commonly used for music-theoretical analysis in the past, the digitisation process afterwards is labour-intensive and error-prone. Accordingly, for carrying out larger annotation experiments on musical patterns, digital tools supporting these annotations need to be developed. This paper analyses the pattern datasets gathered with two recently developed digital annotation tools, ANOMIC (Wells et al. 2019) and PAF (Pesek et al. 2019), using the same musical pieces as in the HEMAN dataset. While these tools overcome the problems of handwritten annotations, they can influence the annotation process in different ways, such as through different music visualisations. We explore the influence of different annotation interfaces and instructions on the discovered patterns, and study differences and commonalities between patterns discovered by annotators of different musical expertise and from different music theoretic schools. We first explore the gathered annotation datasets by performing inter-annotator agreement analysis (Section 4.1), and then employ in a second step a feature analysis of the patterns (Section 4.2) by looking into differences between feature distributions of musical patterns in both datasets. This feature analysis allows more detailed insight into the differences between the pattern datasets. Finally, we discuss the implications of the tools and annotators' backgrounds on the annotation process, and the implications regarding future evaluation methods of pattern discovery algorithms based on rich annotation datasets.

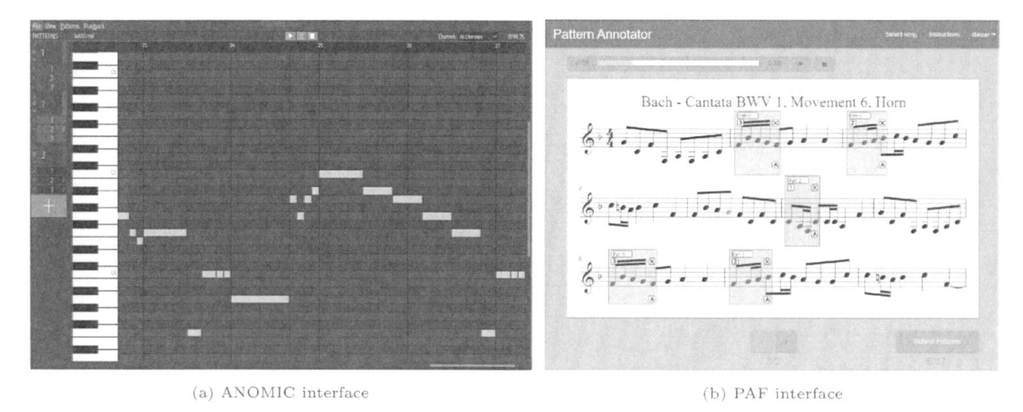

(a) ANOMIC interface

(b) PAF interface

Figure 1. Examples of annotation sessions in the ANOMIC and PAF tools. ANOMIC uses a piano roll representation, PAF uses sheet music representation.

2. Materials and procedure

In this section, we review two digital pattern annotation tools that were developed concurrently – ANOMIC and PAF – and two annotation experiments conducted separately using these tools. While both tools allow for easier digital annotation of patterns, they differ in their implementation, functionality, interface, and music representation.

2.1. *Tools*

The **ANOMIC** (AnNOtate MusIC) tool for pattern annotation was developed as a standalone application for the Windows operating system (Wells et al. 2019). The main view of the tool visualises the MIDI representation of a music piece as a piano roll by plotting the musical notes as rectangles on a two-dimensional onset-time – pitch canvas. This approach is also commonly used in music editing software, as it allows for easy interaction with MIDI elements (most commonly musical notes). Figure 1(a) shows ANOMIC's interface, where the piano roll representation is visible.

The **PAF** (Pattern Annotation Framework) tool[1] was developed as an online application (Pesek et al. 2019) and it visualises the sheet music of a selected piece of music, as seen in Figure 1(b). Thus, it is designed to be used mainly by individuals with musical expertise in order to acquire insightful annotations that are based on musical knowledge. By open-sourcing the tool,[2] the authors hope to aid other researchers in the MIR field who are dealing with pattern-related data gathering and to invite them to contribute additional features to the tool.

Differences between the tools: The most glaring difference between the tools concerns the different visualisations of music they employ: piano roll or sheet music. These different visualisations can influence the users' annotation process. For example, the sheet music representation of PAF is more compact, since notes take up less space than bars. Thus, it is possible to fit larger sections of a music piece onto the screen.

The tools also differ in how they enable the selection of patterns and the annotation of their occurrences. ANOMIC offers click-and-drag actions to select patterns, while annotating with the PAF tool is done by clicking on the starting and then ending note of a pattern. The click-and-drag approach could be perceived as more intuitive, especially considering the left-to-right piano roll visualisation. When the annotation starts, for both tools, a default pattern number (ID) is given

[1] Tool available at http://framework.musiclab.si

[2] Source code available at https://bitbucket.org/ul-fri-lgm/patternannotationframework

Table 1. Summary of main differences between the ANOMIC and PAF annotation tools and user backgrounds.

Tool	Platform	Visualisation	Collecting user background	Helper functions
ANOMIC	Windows	Piano roll	External survey: self-rated music skills	Auto-tagging exact repetitions and chromatic transpositions
PAF	Online	Sheet music	Online registration: music study programmes	None

at first, and if the users proceed to a different group of pattern occurrences, they must use a new number to signify this new group.

ANOMIC also enables users to automatically annotate exact repetitions and chromatic transpositions of already annotated patterns, as implemented by an automatic occurrence matching function (Wells et al. 2019). This function may ease the labour-intensive search necessary in order to annotate all occurrences of a pattern, which is non-trivial for annotators (see Volk and Van Kranenburg 2012). While PAF was later updated to include such an automatic annotation feature, it was not available in the version used for gathering the data for the current analysis.

The tools take different data gathering approaches. ANOMIC is a standalone tool and can be used offline. The gathered patterns must then be submitted via external means, which can affect the number of gathered submissions. Meanwhile, PAF is implemented as an online tool, which automatically saves the annotated patterns in a database. In addition, users must first register and provide their background information to access the tool. In comparison, the ANOMIC tool does not come with built-in registration or user information gathering functionality, so researchers must resort to online surveys. The tools also differ in the overall annotation process: PAF does not allow re-annotation of music pieces and only allows annotation of the provided music pieces, while ANOMIC allows annotation of any music piece the user provides. We summarise the main differences between these two tools in Table 1.

2.2. *Data*

Both tools were used to conduct separate pattern gathering experiments on six monophonic music excerpts from the experiment in Nieto and Farbood (2012), which have previously been used for gathering the HEMAN dataset by Ren, Koops, et al. (2018):

(1) Bach–Cantata BWV 1, Mv. 6, Horn (20 bars)
(2) Bach–Cantata BWV 2, Mv. 6, Soprano (15 bars)
(3) Beethoven–Sq., Op.18, No.1, Violin I (60 bars)
(4) Haydn–Sq., Op.74, No.1, Violin I (30 bars)
(5) Mozart–Sq., K.155, Violin I (28 bars)
(6) Mozart–Sq., K.458, Violin I (54 bars)

These pieces have been selected by Nieto and Farbood (2012) for pattern annotation experiments due to their differing musical characteristics. For example, the Bach chorale is short and has very little rhythmic variation, while the Beethoven string quartet is long, and has clear repetitions varied in pitch. For long pieces, we use roughly the first page of sheet music (precise bar numbers are included in the listing).

2.3. *Procedure: instructions*

For the PAF tool, the instructions for the participants closely followed the ones used in Nieto and Farbood (2012) for musical experts:

> "A musical pattern or motive is defined as a short musical idea, a salient recurring figure, musical fragment, or succession of notes that has some special importance in or is characteristic of a composition. It should not be longer than a musical phrase. If you find a motive that is similar to another (or multiple versions of a motive), choose the one that you think is the most representative."

For ANOMIC, a demo video for the experiment was provided and the instructions on the pattern annotation, accessible to users with little musical training, were as follows:

> "Patterns are distinct, short musical segments or phrases that are considered to be characteristic to a given piece of music and appear multiple times throughout the piece. Be sure to listen to the music and annotate these patterns and their occurrences using the tool. The occurrences don't need to be exact matches, but they should be closely related (compare this to finding occurrences of a leitmotif in a film soundtrack, for example)."

The description of a musical pattern is similar across the two experiments. There are two marked differences between the instructions: whether participants were asked to listen to the music, and how to annotate occurrences of a pattern. While the instructions of ANOMIC explicitly mentioned the importance of occurrences, PAF users were instructed to choose a representative pattern when they see multiple similar ones. ANOMIC users were explicitly asked to listen to the music, whereas PAF users were not. In our analysis in Section 4, we will treat these differences in parallel with the differences of the tools. To be clear about our terminology, henceforth, by "annotations," we mean the data entries created by the annotators; by "occurrences," we mean the repetitions of patterns.

2.4. *Procedure: data collection experiments and participants*

Throughout the two experiments, 1155 pattern annotations were gathered from 39 annotators. 26 of the annotators were involved in the ANOMIC experiment and 13 in the PAF experiment. General information about the annotators (e.g. background surveys regarding musical expertise) was also gathered, thus enabling analysis of relations between the users' backgrounds and the annotations.

In the **ANOMIC** experiment, users installed the tool on their computers and were then asked to submit their annotations via email, along with a completed survey about their background. In total, 788 patterns[3] were gathered, annotated by 26 participants with diverse backgrounds. The participants' levels of musical expertise were assessed through a survey including 10 questions (e.g. ability to read sheet music, proficiency in playing an instrument, academic degree in music), leading to a score between 1 and 10. The annotators were then split into two groups with the cutoff at 5, termed the musicians (14 participants) and the non-musicians group (12 participants), as a rough approximation of their musical expertise (see Wells et al. 2019 for all details). The average scores are 6.71 with a standard deviation of 1.98 for the musicians and 2.42 with standard deviation of 1.31 for the non-musicians group. In future experiments, more sophisticated musical expertise indexes could be used, following Müllensiefen et al. (2014).

The **PAF** tool was used by experts only, namely 13 students attending three faculty study programmes: 4 students from the musicology masters programme at the Faculty of Arts, University of Ljubljana (in the following termed as MU), 3 students of the Music Academy, including music theory and composition at the University of Ljubljana (termed as TC), and 6 students from the music pedagogy programme at the Faculty of Education, University of Maribor (termed as PE). The registration process collected the above information regarding the users' musical backgrounds. Once logged in users were presented with a description of the tool and a summary of its features. In total, 367 annotations were gathered. The majority of participants had between 5–10 years of instrument experience and between 8–15 years of music theory experience.

---

[3] Available at https://github.com/StephanWells/ANOMIC-dataset

3. Methods

There are two main exploratory aspects of our analysis: differences between annotators with different backgrounds and differences between annotations gathered with two different tools. Both will be explored by analysing inter-annotator agreement and by analysing the distributions of various pattern features, for gaining deeper insight as to why annotators might have disagreed. In the following, we describe these methods in more detail.

3.1. *Inter-annotator agreement*

We adopt the methodology for computing inter-annotator agreement from Ren, Koops, et al. (2018) and provide a summary below. An important concept for computing agreement is that of "matched" annotations. Given two pattern occurrences P_1 and P_2, with beginnings and endings denoted as b_1, b_2 and e_1, e_2, were considered to be matched when $|b_1 - b_2| + |e_1 - e_2| \leq T$, where T denotes a threshold value. The vertical bar notation indicates "taking absolute value" to disambiguate from taking cardinality of sets.

Given two sets of pattern occurrences R and C from two annotators, we call one set the reference (R). It does not matter which set is taken as the reference, because we will eventually consider the other set as the reference as well. Using # as "the number of" sign, we then calculate the commonly used measures, namely, precision (#matched_annotations/#R), recall (#matched_annotations/#C), and F_1 score (the harmonic mean of recall and precision) for all possible pairings of annotators. Each individual annotator is compared to every other annotator.

With a number of annotations in a piece, we can expect the precision, recall, and F_1 score to give us a summary of the agreed patterns between any pairs of annotators. These measures will fit the intuition that the more far apart the different annotated pattern boundaries are, the more they disagree; the greater the number of patterns the annotators disagreed upon, the more two annotators disagree. For example, if annotator A noted that the second bar of a musical piece is a pattern, while annotator B included the last quaver of the first bar and the second bar as a pattern, we have the same pattern ending, but a slightly different beginning. The threshold value gives us the flexibility to configure whether the two annotators should be considered to be in agreement (matched) or not. In the example above, if the two patterns are the only annotations in the piece, we have an F_1 score of 0 if $T < 1$ quaver, 1 otherwise. In this way, we can see how much disagreement there is on different scales of time resolution. The reason that we focus on the beginnings and endings of patterns is that, within the same piece of music, once the beginnings and endings are determined, the content of the excerpt is the same for monophonic melodies.

In the following analysis the starts and ends were measured in crotchets and the threshold was set to 5 crotchets as default for this paper, following Ren, Koops, et al. (2018). We will also see that a threshold of 1 crotchet was used for comparison later on. In future work, other threshold values or dynamic thresholds should be investigated.

We will not make a comprehensive cross-comparison between ANOMIC and PAF using this measure, because of the single- or multiple-occurrence difference in the instructions of the tools. The concept of "matching annotations" is a complicated one if we compare the most representative occurrence annotation of a pattern with all the occurrences annotated for a pattern. In addition, "the most representative" and "all occurrences" are not guaranteed as the annotators can only do the best they can. We will, therefore, leave this to be explored in future work.

3.2. *Feature-based annotation comparison*

In order to further explore the differences between annotations of the two digital tools, we compared both annotation datasets on 33 different pattern features, of which we report 7 here, for

Table 2. Descriptions of pattern features analysed in this paper.

Pattern Feature	Feature description
Pattern Duration	The duration of a pattern in crotchets
Occurrences	The number of times that a pattern occurs
Last Note Duration	The duration of the last note of a pattern
Note Range	The number of semi-tones between the lowest and highest note of a pattern
Pitch Direction Changes	The number of melodic arcs in a pattern
Intervallic Leaps	The fraction of all intervals of a pattern that are larger than two semitones
Root Notes	The fraction of notes in a pattern that are root notes or their octaves

simplicity, while all 33 musical features are described in detail online.[4] Most of the features were inspired by the work of Collins (2011), in which musical patterns were rated based on a myriad of musical features from past research, including Meredith, Lemström, and Wiggins (2002), Pearce and Wiggins (2007), Conklin and Bergeron (2008), Forth and Wiggins (2009), as well as Cambouropoulos (2006) and others. Several features were also inspired by the research of Van Kranenburg, Volk, and Wiering (2013), in which global and local features of folk song melodies were compared.

To compare features of the patterns, we take the first occurrence of each pattern annotated in ANOMIC, as PAF annotators only annotated the most representative occurrence of each pattern. We take the first occurrence because the first occurrence of a pattern in a musical piece tends to have a more significant role according to Schoenberg (1967). This taking-the-first-occurrence approach has an exception for one feature, the Occurrence feature, where we use all occurrences from ANOMIC to see whether the annotators actually followed the instructions closely regarding annotating single or multiple occurrences.

We then computed musical features of each pattern, thus forming feature distributions for both datasets. Next, the distributions of each feature were normalised by taking the minimum and maximum values across both distributions and performing min-max scaling. The computation process of each feature is described in detail in the supplemental online material,[4] which also includes the Python source code used for the analysis.

In Table 2, we list the 7 features included in this paper, namely those which we considered to be most intuitively related to pattern characteristics perceivable by users (such as the duration of the last note or the note range). Notice that the Occurrence feature is what we mentioned as an exception to the taking-the-first-occurrence approach above. Furthermore, following the comparison of local and global features in Van Kranenburg, Volk, and Wiering (2013), we focussed on local features, which are more likely to be assessed by humans when annotating patterns. We further reduced the number of important features by analysing the Spearman's and Pearson correlation coefficients between pairs of features. The highest Pearson correlation value appeared between the feature pattern duration and note range, which has a correlation of 0.64 and 0.65 for ANOMIC and PAF. Tables of the most correlated features for each dataset can be accessed online.[4]

Our last step to analyse pattern features is based on independent two-sample Student tests (t-tests), two sample Kolmogorov-Smirnov tests (KS-tests) and boxplot visualisations of distributions. Given our null hypothesis that the samples are drawn from the same distribution, and that we are unsure whether our data is normally distributed, we use both parametric and nonparametric tests for verifying how likely it is that the distributions actually differ. The differences

[4] Results available at https://bitbucket.org/dariant1/agreement-in-musical-patterns

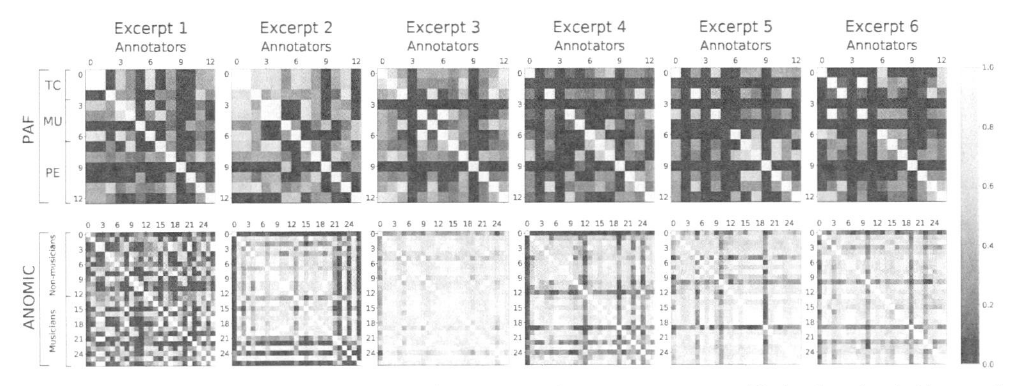

Figure 2. F_1 score matrices, representing pairwise agreement between annotators, with the time threshold set to 5 crotchets. Each matrix showcases results for a different music excerpt. Matrix columns and rows denote different annotators, grouped based on the annotation tool used (PAF or ANOMIC) and their musical background. Matrix cell colours (These figures are best viewed in an online version of the paper.) correspond to the obtained pairwise agreement values, where yellow denotes high agreement and blue indicates low. Some annotators did not provide annotations for all excerpts, which can be seen along the diagonals as low agreement.

are represented by high t and D statistic values in combination with low p values of the performed tests. We also considered boxplot visualisations of the distributions to better understand the shapes of the distributions and the differences between them.

4. Results

4.1. *Inter-annotator agreement*

In order to analyse inter-annotator agreement we computed the F_1 scores of all annotator pairs across all music pieces. We gathered these values in F_1 score matrices, which allow for a concise presentation of results. Figure 2 shows these matrices, in which annotators are grouped based on the annotation tool used and their backgrounds. The analysed groups include the TC, MU and PE groups of PAF as well as the musician and the non-musician groups of ANOMIC (see Section 2.4). Based on the obtained inter-annotator agreement values, we refer to the values above 0.95 (yellow matrix values) as indicators for a strong agreement in this paper. It should be noted that the number of annotations was not split equally among music excerpts, as some annotators of the PAF tool did not annotate the last three excerpts. This was likely due to the fixed order of music excerpts and the significant time investment in the annotation process. Once this issue of the PAF tool was identified, it was reported and addressed by the developers who randomised the ordering to improve the tool for future use.

The TC and MU groups show strong agreement on the first three music excerpts. Meanwhile, the results of the PE group show many weaker agreements, despite a larger number of annotators. Contrary to the TC and MU groups, the annotators of the PE group did somewhat agree in the annotations for excerpt 4. There is only one strong agreement between annotators belonging to different groups (excerpt 6, annotators 1 and 4). Since there are several strong agreements between the annotators within individual groups, the lack of agreement between different groups could indicate the potential influence of different study programmes on the annotators' perception of the most representative musical patterns.

For ANOMIC, the agreement values for the two subgroups are similar: the average inter-annotator agreement for the musician group is 0.61, and 0.63 for non-musicians. While some agreement does exist between the ANOMIC and PAF annotators, we only observe one strong agreement between two PAF annotators and one non-musician annotator of ANOMIC in excerpt 1. We do not make further comparisons between PAF and ANOMIC disagreements, as a range of factors could have contributed to their differences, such as differences in instructions and the threshold value.

Next, we lower the threshold for agreement computation from 5 crotchets to 1 to see what changes may be brought on to our results by a different threshold value. The agreement values become much smaller among the ANOMIC non-musicians (0.38) and the musicians (0.47), for the crotchet threshold of 1. For the same threshold the average agreement between all ANOMIC participants was 0.40.

We have an additional note for these comparisons. In taking averages, we can compare between groups while marginalising the effects of individual differences between musical pieces. We are aware that this is not always valid because there is a varying degree of difficulty in finding patterns across different musical pieces. It is possible that a group of annotators disagree strongly on one single piece and agree perfectly with each other on the rest, which would be obscured in the average, with the music being a confounding factor. However, in Figure 2, we see a range of disagreement and agreement. Admittedly, excerpt 1 is more disagreed upon than others, so we also calculated the values by only using the other five excerpts, and the results and conclusions did not change. Furthermore, the computation of the average was based on the whole matrices, thus including values where users did not provide any annotations. These values were simply set to 0 and were included in the computation. We also analysed the average values if these values were ignored. Despite affecting the average values of the comparison, the changes in values were not significant since the values, based on PAF users, simply increased by around 0.01. Thus, we decided to only report the original values, which included missing annotations.

4.2. *Feature-based comparison of annotation datasets*

As introduced in Section 3, we analysed, for each pattern feature, the annotations of ANOMIC annotators with musical and non-musical backgrounds and compared them to the PAF dataset, whose annotators all had a musical background. We investigated whether the difference between datasets was also present in these background subgroups to identify if the observed difference between the PAF and the ANOMIC dataset was influenced primarily by the tools or the musical backgrounds of the annotators.

The results of the analysis revealed 23 out of the 33 features, where significant differences were seen between the annotations of the PAF and the ANOMIC tool. We then eliminated several features, based on how intuitive they are and their correlations, and narrowed the list down to 7 features: pattern duration, occurrences, last note duration, note range, pitch direction changes, intervallic leaps and root notes. These features are also listed in Table 2, along with their descriptions. Plotted distributions of these features can be seen in Figure 3. The calculated t and p values of the t-tests between the selected features are shown in Table 3. A list of all analysed features, all t-test and KS-test results as well as all boxplot visualisations are available online.[4]

4.2.1. *Duration and occurrence features*

The first of the features we analysed was the **Pattern Duration** feature, which measures the length of a pattern in crotchets (quarter notes). We observed that the ANOMIC distribution had much smaller overall and interquartile ranges than the PAF distribution. The mean and median values of the distribution were also much smaller.

Next was the **Occurrences** feature, which refers to the number of times that a pattern occurs in a music excerpt, as defined by Collins (2011). From the boxplots in Figure 3 it is evident that

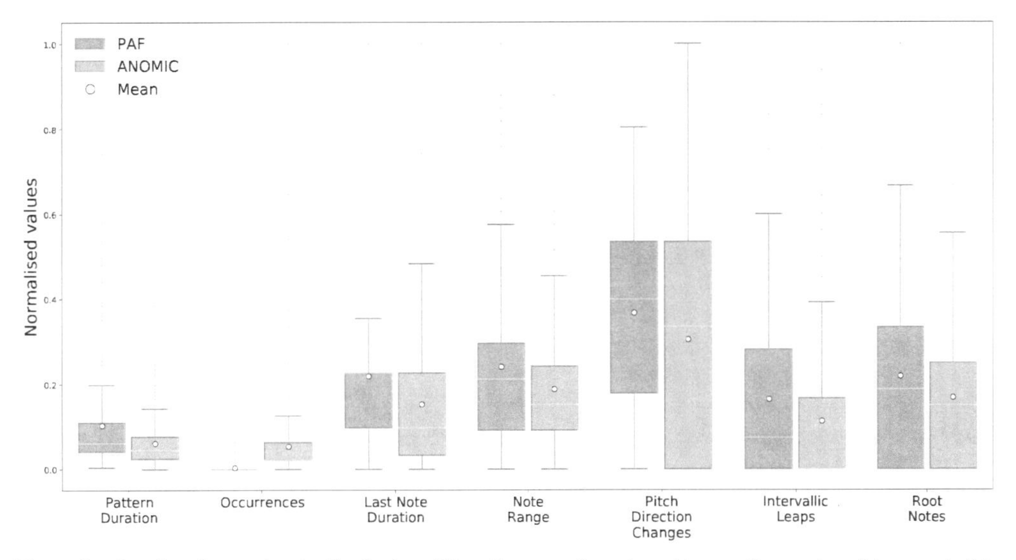

Figure 3. Boxplots showcasing the distributions (These figures are best viewed in an online version of the paper.) of the analysed pattern features of the PAF and the ANOMIC datasets. For each feature the two distributions were normalised.

Table 3. Table showing t and p values of t-tests between the PAF and the ANOMIC (Mus. and Non-Mus.) datasets on the features of the first column.

	t-tests between pattern features (t and p values of t-test)			
	PAF/ANOMIC	PAF/Mus.	PAF/Non-Mus.	Mus./Non-Mus.
Pattern Duration	$7.54\ (9.60 \times 10^{-14})$	$4.06\ (5.38 \times 10^{-05})$	$7.68\ (5.09 \times 10^{-14})$	$4.91\ (1.09 \times 10^{-06})$
Occurrences	$-11.96\ (4.35 \times 10^{-31})$	$-16.42\ (2.98 \times 10^{-51})$	$-11.12\ (1.13 \times 10^{-26})$	$-3.1\ (1.98 \times 10^{-03})$
Last Note Dur.	$6.75\ (2.45 \times 10^{-11})$	$3.93\ (9.30 \times 10^{-05})$	$7.36\ (4.90 \times 10^{-13})$	$3.49\ (5.15 \times 10^{\ 04})$
Note Range	$5.18\ (2.57 \times 10^{-07})$	$\mathbf{1.62\ (1.05 \times 10^{-01})}$	$6.87\ (1.36 \times 10^{-11})$	$6.15\ (1.23 \times 10^{-09})$
Pitch Direct. Ch.	$3.53\ (4.25 \times 10^{-04})$	$\mathbf{-0.02\ (9.86 \times 10^{-01})}$	$5.98\ (3.55 \times 10^{-09})$	$5.92\ (4.79 \times 10^{-09})$
Intervallic Leaps	$3.89\ (1.07 \times 10^{-04})$	$\mathbf{1.95\ (5.11 \times 10^{-02})}$	$4.7\ (3.07 \times 10^{-06})$	$2.75\ (6.12 \times 10^{-03})$
Root Notes	$4.45\ (9.26 \times 10^{-06})$	$3.51\ (4.73 \times 10^{-04})$	$3.95\ (8.67 \times 10^{-05})$	$\mathbf{0.37\ (7.08 \times 10^{-01})}$

Note: Orange cells denote tests with $p > 0.05$, which means that the difference between the two tested distributions is not statistically significant.

the PAF distribution differs drastically from the ANOMIC distribution, which is expected due to the difference in instructions give about whether to annotate all occurrences or not. This result confirms that the annotators followed the instructions closely in this respect.

Differences between the two datasets were also seen for the **Last Note Duration** feature, which describes the duration of the last note of a pattern. Here the ANOMIC distribution had larger interquartile and overall ranges than the PAF distribution, while its mean and median values were lower.

Results of the t-tests in Table 3 and the KS-tests (available online)[4] confirmed that the above mentioned differences between distribution pairs were statistically significant. Furthermore, they revealed that the differences between the PAF dataset and the two subgroups of the ANOMIC dataset, based on musical backgrounds, were also statistically significant. Thus, we may conclude that the differences between feature distributions of pattern datasets were caused by varying musical background but also possibly by the interfaces or the instructions of the annotation tools.

4.2.2. *Pitch-based features*

The two annotation datasets also differed on the **Note Range** feature, which describes the number of semitones between the lowest and highest note of a pattern. Figure 3 shows that the PAF distribution has a much larger overall and interquartile range than the ANOMIC distribution. Its median and mean values are also higher.

Similar differences were also seen for the **Pitch Direction Changes** feature, which is defined as the number of melodic arcs in a pattern (inspired by various melodic arc features in Van Kranenburg, Volk, and Wiering 2013). Our analysis showed that the ANOMIC distribution had a much larger overall and interquartile range. Furthermore, the distribution lacks the bottom whisker, thus being more positively skewed. The ANOMIC distribution also had significantly lower mean and median values.

We observed comparable differences among the **Intervallic Leaps** feature distributions. The feature describes the fraction of all melodic intervals of a pattern whose note range is larger than two semitones, as defined by Collins (2011). Considering the analysis results, we see that the PAF distribution has much larger overall and interquartile ranges. Its median and mean values are also significantly higher.

Finally, differences were seen for the **Root Notes** feature, which describes the fraction of notes in a pattern that are root notes or octaves of roots of the music piece. From the distributions present in Figure 3, we can discern that the distributions mainly differ in the overall and interquartile ranges, with the PAF distribution having much larger ranges than the ANOMIC distribution. We also note that the mean value as well as the standard deviation of the PAF distribution are slightly higher that those of ANOMIC.

By analysing the t-test results from Table 3 and the KS-test results (available online),[4] we confirm that the distribution differences of the first three pitch-based features between the two datasets are statistically significant. We notice that these three features pertain to the relationships between notes in the patterns and are therefore melodically relevant. We can also discern that the differences are not statistically significant between the PAF group and the musician group of ANOMIC. However, the difference between musicians, both PAF and ANOMIC, and non-musicians of ANOMIC, is statistically significant.

For the **Root Notes** feature, the analysis shows that the difference between the two ANOMIC subgroup distributions is not statistically significant. Thus, they are considered to be similar. Meanwhile, the difference between the PAF distribution and both subgroup distributions is statistically significant.

5. Discussion

The analysis of the inter-annotator agreement and the feature-based comparison of the annotation datasets have revealed differences in the annotated patterns between user groups and between annotation tools and their instructions. The agreement results show that musical background has an influence on the annotations; annotators attending the same programme tend to agree more than others.

Several differences observed in the feature-based comparison of the annotation datasets point towards discrepancies in annotations caused by design differences in the tools and experiments. The PAF interface displays sheet music allowing a compact representation, with large sections of the music piece being presented to the user within a single view. By contrast, ANOMIC's piano roll representation generally displays fewer elements at once to the user in order to preserve element clarity. The difference in music visualisation might have caused users to perceive and annotate patterns with durations relative to the view window size. The notes in the sheet music representation of PAF remained roughly the same size, while the piano roll elements of ANOMIC varied in their size based on the durations. Since the last note durations of the PAF dataset are on average significantly longer, differently sized elements of the ANOMIC tool could have discouraged users from picking longer notes as pattern endings. The drastic difference between the occurrences feature distributions was likely caused by the lack of an automatic occurrence matching functionality in the PAF tool and by the difference in the instructions given to the users.

Moreover, we observed differences in note range, intervallic leaps, and pitch direction changes between ANOMIC non-musicians versus ANOMIC musicians and PAF annotators. The t-test analysis showed that the differences were not statistically significant between the PAF group and the musician group of ANOMIC. However, the difference between musicians, both PAF and ANOMIC, and non-musicians of ANOMIC, was statistically significant. We therefore conclude the varying musical background of the ANOMIC annotators as the underlying cause, and exclude the potential influence of the annotation tools. Finally, for the root notes feature we are unsure what might have caused the difference in distributions, though we believe it is not caused by the musical background based on the lack of difference between feature distributions of the ANOMIC musicians and non-musicians.

In this first explorative study of comparing annotation tools, we did not streamline instructions for musical experts (PAF) with instructions also addressing non-musicians (ANOMIC). Including more specific instructions in the future may enable more controlled conditions when comparing different tools. For instance, an instruction to first listen to the music without consulting the visualisation before starting with the annotation process, might decrease an otherwise perhaps strong tendency of users to annotate patterns they can visually identify. Nevertheless, each music visualisation will have an influence on the annotation process to a certain extent. Musical experts may be most familiar with sheet music, but piano roll visualisations may be more accessible for annotators with less musical expertise. As our results indicate that the size of the musical excerpt that can be displayed in a single view to the user seems to influence the length of patterns annotated, this should be specifically considered when longer patterns are expected to be important for a specific corpus.

Moreover, the influence of the automatic occurrence matching functionality needs to be investigated in more depth in the future. If the goal of the annotation is to find all occurrences of a given pattern, as was the case in the ANOMIC experiment, it can alleviate the finding of the pattern occurrences for users, but might have the side effect of pointing users to occurrences they would not have deemed important otherwise. If only the most representative pattern should be annotated, as in the PAF experiment, such a tool can assist in highlighting all occurrences from which the user can then choose the most representative one for the annotation. More inexact repetition matching functions may be added so that the found pattern occurrences are not biased towards exact repetitions or chromatic transpositions. Either way, this calls for a systematic investigation of using tools with and without such functionality.

6. Conclusion

In this study, we compared two digital pattern annotation tools, PAF and ANOMIC, and analysed two pattern datasets collected with these tools on the same musical pieces, employing inter-annotator agreement analysis and feature-based analysis of musical patterns. Comparing the annotations collected with the PAF tool enabled us to study three groups of annotators from different musical study programmes. We observed higher agreement between annotators of the same group when compared to annotators of different groups, indicating a potential influence of study programmes on the understanding and perception of patterns in music. Comparing the annotations of the ANOMIC annotation set gave a similar result.

Our findings point to a major influence of the annotation tools, instructions, and the musical background of participants on the annotated patterns. As a next step, the influence of the tools should be studied in more detail using stricter controlled comparisons, including a clarification on how users should include their listening experience into the annotation process, and a controlled use of the automatic pattern matching functionality. Moreover, the analysis of the pattern datasets can be enriched by further investigations as to where annotators tend to agree, for instance by exploring dynamic thresholds for calculating inter-annotator agreement depending on the size of the patterns. Determining in more detail different levels of granularity as to when two pattern annotations can be considered as agreeing, even if the exact beginning and ending points are not identical, can further help to identify different layers of commonality between annotators.

We believe that the widespread use of digital tools in gathering pattern annotations is inevitable in the near future. Our findings point to several directions for improvements of large-scale data collection and analysis of musical patterns. The observed differences in annotations gathered with different tools call for further experiments and analyses for deriving technical design choices that fit the purpose of pattern annotations in an optimal way for a given context and annotator group. Establishing reference data for evaluating automatic pattern discovery algorithms from such rich annotation datasets can follow different directions. For instance, identifying subgroups of annotators that highly agree with each other can assist in establishing single-reference data based on a larger group of annotators. Establishing evaluation methods that take into account multiple reference annotations expressing different subjective interpretations of the same musical piece, can pave the way for a more adequate consideration of ambiguity and subjectivity in the evaluation of pattern discovery algorithms.

Acknowledgement

We would like to thank the editors Darrell Conklin and Jason Yust, as well as the anonymous reviewers, for their constructive and helpful comments during the revision process.

Disclosure statement

No potential conflict of interest was reported by the author(s).

Supplemental online material

Supplemental data for this article can be accessed online at https://zenodo.org/record/4544002#.YO7bOD3iuUk.

References

Balke, S., J. Driedger, J. Abeßer, C. Dittmar, and M. Müller. 2016. "Towards Evaluating Multiple Predominant Melody Annotations in Jazz Recordings." In *Proceedings of the International Conference on Music Information Retrieval (ISMIR)*, 246–252. New York City, United States of America.

Bamberger, J. S. 2000. *Developing Musical Intuitions: A Project-Based Introduction to Making and Understanding Music*. New York City, United States of America: Oxford University Press.

Boot, P., A. Volk, and W. B. de Haas. 2016. "Evaluating the Role of Repeated Patterns in Folk Song Classification and Compression." *Journal of New Music Research* 45 (3): 223–238.

Cambouropoulos, E. 2006. "Musical Parallelism and Melodic Segmentation: a Computational Approach." *Music Perception* 23 (3): 249–268.

Collins, T. 2019. "Discovery of Repeated Themes & Sections – MIREX Wiki." `https://www.music-ir.org/mirex/wiki/2019:Discovery_of_Repeated_Themes_%26_Sections`.

Collins, T. E. 2011. "Improved Methods for Pattern Discovery in Music, with Applications in Automated Stylistic Composition." PhD thesis, The Open University.

Conklin, D., and M. Bergeron. 2008. "Feature Set Patterns in Music." *Computer Music Journal* 32 (1): 60–70.

Flexer, A., and T. Grill. 2016. "The Problem of Limited Inter-Rater Agreement in Modelling Music Similarity." *Journal of New Music Research* 45 (3): 239–251.

Forth, J., and G. A. Wiggins. 2009. "An Approach for Identifying Salient Repetition in Multidimensional Representations of Polyphonic Music." In *London Algorithmics 2008: Theory and practice, Texts in Algorithmics*, 44–58. London, United Kingdom: College Publications.

Herremans, D., and E. Chew. 2017. "MorpheuS: Generating Structured Music with Constrained Patterns and Tension." *IEEE Transactions on Affective Computing* 10 (4): 510–523.

Janssen, B., W. B. De Haas, A. Volk, and P. Van Kranenburg. 2013. "Finding Repeated Patterns in Music: State of Knowledge, Challenges, Perspectives." In *International Symposium on Computer Music Multidisciplinary Research*, 277–297. Springer.

Koops, H. V., W. B. de Haas, J. A. Burgoyne, J. Bransen, A. Kent-Muller, and A. Volk. 2019. "Annotator Subjectivity in Harmony Annotations of Popular Music." *Journal of New Music Research* 48 (3): 232–252.

Margulis, E. H. 2014. *On Repeat: How Music Plays the Mind.* New York City, United States of America: Oxford University Press.

Melkonian, O., I. Y. Ren, W. Swierstra, and A. Volk. 2019. "What Constitutes a Musical Pattern?" In *Proceedings of the 7th ACM SIGPLAN International Workshop on Functional Art, Music, Modeling, and Design*, 95–105. Berlin, Germany.

Meredith, D., K. Lemström, and G. A. Wiggins. 2002. "Algorithms for Discovering Repeated Patterns in Multidimensional Representations of Polyphonic Music." *Journal of New Music Research* 31 (4): 321–345.

Müllensiefen, D., B. Gingras, J. Musil, and L. Stewart. 2014. "The Musicality of Non-Musicians: An Index for Assessing Musical Sophistication in the General Population." *PLOS One* 9 (2): e89642.

Nieto, O., and M. M. Farbood. 2012. "Perceptual Evaluation of Automatically Extracted Musical Motives." In *Proceedings of the 12th International Conference on Music Perception and Cognition*, 723–727. Thessaloniki, Greece.

Pearce, M. T., and G. A. Wiggins. 2007. "Evaluating Cognitive Models of Musical Composition." In *Proceedings of the 4th International Joint Workshop on Computational Creativity*, 73–80. London, United Kingdom.

Pesek, M., D. Tomašević, I. Y. Ren, and M. Marolt. 2019. "An Opensource Web-Based Pattern Annotation Framework – PAF." In *Proceedings of the International Conference on Music Information Retrieval (ISMIR)*, 1–2. Delft, The Netherlands.

Ren, I. Y., H. V. Koops, A. Volk, and W. Swierstra. 2018. "Investigating Musical Pattern Ambiguity in a Human Annotated Dataset." In *Proceedings of the 15th International Conference on Music Perception and Cognition and the 10th triennial conference of the European Society for the Cognitive Sciences of Music*, 361–367. Graz, Austria.

Ren, I. Y., A. Volk, W. Swierstra, and R. C. Veltkamp. 2018. "Analysis by Classification: A Comparative Study of Annotated and Algorithmically Extracted Patterns in Symbolic Music Data." In *Proceedings of the International Conference on Music Information Retrieval (ISMIR)*, 539–546. Paris, France.

Schoenberg, A. 1967. *Fundamentals of Musical Composition*. London: Faber and Faber.

Sears, D. R. W., and Gerhard Widmer. 2020. "Beneath (Or Beyond) the Surface: Discovering Voice-Leading Patterns with Skip-Grams." *Journal of Mathematics and Music* 1–26. https://www.tandfonline.com/doi/full/10.1080/17459737.2020.1785568

Taube, H. 1995. "An Object-Oriented Representation for Musical Pattern Definition." *Journal of New Music Research* 24 (2): 121–129. https://doi.org/10.1080/09298219508570678.

Van Kranenburg, P., A. Volk, and F. Wiering. 2013. "A Comparison Between Global and Local Features for Computational Classification of Folk Song Melodies." *Journal of New Music Research* 42 (1): 1–18.

Volk, A., and P. Van Kranenburg. 2012. "Melodic Similarity Among Folk Songs: An Annotation Study on Similarity-Based Categorization in Music." *Musicae Scientiae* 16 (3): 317–339.

Wells, S., A. Volk, J. Masthoff, and I. Y. Ren. 2019. "Creating a Tool for Facilitating and Researching Human Annotation of Musical Patterns." Master's thesis, Universiteit Utrecht.

Index

Note: Page numbers followed by "n" denote endnotes.